How does our Universe evolve? And how did structures like stars and galaxies form? In recent years, scientists' understanding of these profound questions has developed enormously. This book presents a clear and detailed picture of contemporary cosmology for the general reader.

Unlike existing popular books on cosmology, *After the first three minutes* does not gloss over details, nor shy away from explaining the underlying concepts. Instead, with a lucid and informal style, the author introduces all the relevant background and then carefully pieces together an engaging story of the evolution of our Universe. We are left with a state-of-the-art picture of scientists' current understanding in cosmology, and a keen taste of the excitement of this fast-moving science. Throughout, no mathematics is used; and all technical jargon is clearly introduced and reinforced in a handy glossary at the end of the book.

For general readers who want to get to grips with what we really do and don't know about our Universe, this book provides an exciting and uncompromising read.

T PADMANABHAN

Inter-University Centre for Astronomy and Astrophysics, Pune, India

AFTER
THE FIRST THREE MINUTES
THE STORY OF OUR UNIVERSE

CAMBRIDGE
UNIVERSITY PRESS

PUBLISHED BY THE PRESS SYNDICATE OF THE UNIVERSITY OF CAMBRIDGE
The Pitt Building, Trumpington Street, Cambridge CB2 1RP, United Kingdom

CAMBRIDGE UNIVERSITY PRESS
The Edinburgh Building, Cambridge CB2 2RU, United Kingdom
40 West 20th Street, New York, NY 10011-4211, USA
10 Stamford Road, Oakleigh, Melbourne 3166, Australia

First published 1998

Printed in the United Kingdom at the University Press, Cambridge

Typeset in Ehrhardt 11/14 pt [KW]

A catalogue record for this book is available from the British Library

Library of Congress Cataloguing in Publication data
Padmanabhan, T. (Thanu), 1957-
After the first three minutes : the story of our universe / T. Padmanabhan.
p.   cm.
Includes bibliographical references and index.
ISBN 0 521 62039 2
1. Cosmology. I. Title.
QB981.P243 1998
523.1–dc21   97-11060 CIP

ISBN 0 521 62039 2 hardback
ISBN 0 521 62972 1 paperback

. . . to my daughter Hamsa, who thinks I should be
playing with her instead of writing such books. . .

Contents

Preface

One thing I have learnt in a long life: that all our science, measured against reality, is primitive and childlike – and yet it is the most precious thing we have.   A. EINSTEIN

The subject of cosmology – and our understanding of how structures like galaxies, etc., have formed – have developed considerably in the last two decades or so. Along with this development came an increase in awareness about astronomy and cosmology among the general public, no doubt partly due to the popular press. Given this background, it is certainly desirable to have a book which presents current thinking in the subject of cosmology in a manner understandable to the common reader. This book is intended to provide such a non-mathematical description of this subject to the general reader, at the level of articles in *New Scientist* or *Scientific American*. An average reader of these magazines should have no difficulty with this book.

The book is structured as follows: chapter 1 is a gentle introduction to the panorama in our universe, various structures and length scales. Chapter 2 is a rapid overview of the basic physical concepts needed to understand the rest of the book. I have tried to design this chapter in such a manner as to provide the reader with a solid foundation in various concepts, which (s)he will find useful even while reading any other popular article in physical sciences. Chapter 3, I must confess, is a bit of a digression. It provides the useful survey of astronomical observations which a reader will enjoy, but it is closely connected with chapter 4. Because of the interdependence of chapters 3 and 4, I strongly recommend that the reader return to chapter 3 after reading chapter 4. The fourth chapter describes a gamut of astrophysical structures and processes: stars, stellar evolution, galaxies, clusters of galaxies, etc. find their place here. Chapter 5 is devoted to the big bang model of the universe and a description of the early universe. This lays the foundation for *the* key chapter of the book, viz. chapter 6, which deals with the formation of various structures in the universe. The current thinking in this subject is presented in detail, and I have also provided a critical appraisal of models. Chapter 7 deals with the farthest objects in the universe which astronomers are currently studying, viz. quasars, radio galaxies, etc. The final chapter summarizes the entire book and emphasizes the broad picture available today.

The choice of topics necessarily reflects the bias of the author, but I have consciously tried to increase the 'shelf-life' of the book, in spite of the fact that it deals with a rapidly evolving area. For example, chapters 2, 4, 5 and most of 3 will remain useful for a fair amount of time. Some of the details in chapters 6 and 7 will doubtless change, but I think the broad ideas and concepts will remain useful for a longer time.

The discussion is non-mathematical, but I have not avoided mentioning actual numbers, units, etc. when it is required. This is essential in the discussion of several topics; any attempt to present these results in 'pure English' would have oversimplified matters. In fact, I have tried to avoid oversimplification as much as possible, though I might have cut corners as regards details in a few places. This book is intended for the serious reader who *really* wants to know, and I do expect such a reader to put in some effort to understand the ideas which are presented. I have also avoided telling the reader stories; you will not find any of '... as Prof. Great was driving with his wife to the concert, it suddenly occured to him....' kind of stuff in this book. Ideas and discoveries are more important than individuals, and I have kept the discussion mostly impersonal and non–historical. (You will find some exceptions, which are – of course – intentional.) The publishers and I debated whether to go for a glossy book with lots of colour pictures or to go for a no-nonsense version, and decided on the latter. It was a difficult choice, and I hope the clarity of the contents compensates for the lack of colour.

You will find a fairly detailed glossary at the end of the book, containing the technical terminology introduced in various chapters. Since concepts developed in chapter 2 or 4 (say) may be needed in chapter 7, the glossary will serve the useful purpose of reminding the reader of unfamiliar jargon. I have also included a brief list of books for further reading; this choice is definitely biased and should be merely taken as an indicative sample. Some of these books contain more extensive bibliographies.

This book originated from a comment made by John Gribbin. While reviewing my book *Structure formation in the universe* (CUP, 1993), he said that: 'It would be good to see the author attempting a popular book on the same theme'. Rufus Neal of CUP responded positively to this suggestion, and I hope I haven't disappointed them.

Several others have contributed significantly to the making of this book: This is the second time I have worked with Adam Black of CUP, and it was delightful. The entire TEXing, formatting of figures and proof-reading was done by Vasanthi Padmanabhan, and I am grateful to her for the hours of effort she has put in. Most of the figures were done by Prem Kumar (NCRA) and I thank him for his help. I am grateful to my friends Jasjeet Bagla, Mangala Narlikar, Lakshmi Vivekanand, Ramprakash, Srianand, Srinivasan, Sriramkumar and Vivekanand, who have read earlier drafts of the book and have made useful comments and suggestions. I thank IUCAA for the use of computer facilities.   T. PADMANABHAN

1

Introducing the universe

1.1 Cosmic inventory

Think of a large ship sailing through the ocean carrying a sack of potatoes in its cargo hold. There is a potato bug, inside one of the potatoes, which is trying to understand the nature of the ocean through which the ship is moving. Sir Arthur Eddington, famous British astronomer, once compared man's search for the mysteries of the universe to the activities of the potato bug in the above example. He might have been right as far as the comparison of dimensions went; but he was completely wrong in spirit. The 'potato bugs' – called more respectably astronomers and cosmologists – have definitely learnt a lot about the contents and nature of the Cosmos.

If you glance at the sky on a clear night, you will see a vast collection of glittering stars and – possibly – the Moon and a few planets. Maybe you could also identify some familiar constellations like the Big Bear. This might give you the impression that the universe is made of a collection of stars, spiced with the planets and the Moon. No, far from it; there is a lot more to the universe than meets the naked eye!

Each of the stars you see in the sky is like our Sun, and the collection of all these stars is called the 'Milky Way' galaxy. Telescopes reveal that the universe contains millions of such galaxies – each made of a vast number of stars – separated by enormous distances. Other galaxies are so far away that we cannot see them with the naked eye. So what you see at night is only a tiny drop in the vast sea of the cosmos. To understand and appreciate our universe, it is necessary to first come to grips with the large numbers involved in the cosmological description.

To do this systematically, let us start with our own home planet, Earth. The radius of the Earth is about 6400 kilometers (km, for short), and by modern standards of transport this is a small number. A commercial jetliner can fly around the Earth in about 40 hours. The nearest cosmic object to the Earth is the Moon, which is about 400 000 km away. This distance is about 60 times the radius of the Earth, and Apollo 11 – which landed men on the Moon for the first time – took about four days to cover this distance. As you know, the Moon is a satellite of Earth and orbits around us once in about 30 days. The Earth itself, of course, is moving around the Sun, which is about 150 million km away. Apollo 11 would have taken nearly five years to travel this distance. There are several other planets like the Earth which are orbiting the Sun at varying distances from it. The closest one to the Sun is Mercury, the next one is Venus (which may be familiar to you as the 'Evening Star' in the sky) and the third one is the Earth. After Earth comes Mars (the red planet), Jupiter, Saturn (the one with rings), Uranus, Neptune and Pluto in order of increasing distance. The farthest known planet in the solar system, Pluto, is at a distance which is nearly 40 times as large as the distance between the Earth and the Sun. Thus our planetary system has a radius of about 6000 million km.

The distances are already beginning to be quite large, and it will be convenient if we use bigger units to express these astronomical distances. After all, you measure the length of a pencil in centimeters but give the distance between two cities in kilometers, which is a much larger unit. In the same spirit, a terrestrial unit like the kilometer is inadequate to express cosmic dimensions and it would be nice to have bigger units. One such unit, which is used extensively, is called the 'light year'. In spite of the word 'year', this is a unit of *distance* and *not* of time. One light year is the distance travelled by light in one year. Light can travel nearly 300 000 km in one second, and so a light year is about 10 million million km. Written in full, this would be 10 000 000 000 000 km. For convenience of writing, we use the symbol 10^{13} km to denote this quantity. This symbol stands for the number obtained by multiplying 10 thirteen times or – equivalently – the number with 1 followed by 13 zeros. So we may say that one light year is about 10^{13} km.

This unit, the light year, becomes really useful when we consider still larger scales. As we said before, our Sun is only one among 100 000 million stars which exist together in a galaxy named the 'Milky Way'. Using our short-hand notation, we can say that our galaxy has about 10^{11} stars. The star nearest to our Sun is called 'Proxima Centauri' (which is in the constellation Centaurus) and is about four light years away from us. In other words, a light ray – which can go through our planetary system in about eleven hours – will take about four years to reach the *nearest* star! (Apollo 11 would have taken nearly one million years!). All the stars you see in the sky belong to the Milky Way galaxy and are at different distances from us. Each one of them could be as big and bright as our Sun, and could possibly have a planetary system around it. Sirius, the brightest star in the sky, is about eight light years away; the reddish star, Betelgeuse, in the familiar constellation of Orion, is at a distance of 600 light years, and the Pole star is about 700 light years from us. The light which we receive today from the Pole star was emitted during the fourteenth century!

The entire Milky Way galaxy is about 45 000 light years in extent. To describe the galaxy, we find that even the light year is an inadequately small unit. So astronomers use a still bigger unit called a 'kiloparsec' (kpc, for short). One kiloparsec is about 3000 light years. The radius of a typical galaxy (like ours) is about 15 kpc. Kiloparsec is the most often used unit in mapping the universe.

Just as the Sun is only an average star in the Milky Way, our galaxy itself is only one among a large number of galaxies in the universe. Powerful telescopes reveal that our universe contains more than 100 million galaxies similar to ours. The nearest big galaxy to the Milky Way is called 'Andromeda'. This galaxy is at a distance of about 700 kpc, and is just barely visible to the naked eye as a hazy patch in the constellation Andromeda. The Andromeda galaxy, like the Milky Way, is made of about 10^{11} stars (just to remind you, 10^{11} stands for the number 100 000 000 000 – which is 1 followed by 11 zeros; we shall use this notation repeatedly in this book). But since it is so far away we cannot see the individual stars of Andromeda with our naked eye.

Around the Milky Way and Andromeda there exist about 30 other galaxies. Some of these, called 'Dwarf galaxies', are considerably smaller in size, and each of them contains only about a million stars or so, while a few others are bigger. The Milky Way, Andromeda and these galaxies together form the next unit in the cosmic scale, usually called the 'Local Group of galaxies'. The size of the Local Group is about 1000 kpc, which is called 1 Megaparsec (Mpc, for short).

The assembling of galaxies in the form of groups or clusters is also a general feature of our universe. Astronomers have detected a large number of clusters of galaxies, many of which are much bigger than our Local Group (which contains only about 30 galaxies). Some of the large clusters of galaxies – like the one called the 'Coma cluster' – contain nearly 1000 individual galaxies.

Are there structures still bigger than the cluster of galaxies? Observations suggest that clusters themselves could be forming bigger aggregates called 'superclusters', with sizes of about 30 to 60 Mpc. For example, our Local Group is considered to be a peripheral member of a supercluster called the

'Virgo supercluster'. However, the evidence for superclusters is not as firm as that for clusters.

How big is the universe itself? With powerful telescopes we can now probe distances of about 3000–6000 Mpc. In other words, we can say that the size of the observable region of the universe is nearly 1000 times bigger than the size of a galaxy cluster.

You must have noticed that, in going from the Earth to a supercluster, we have come across a hierarchy of structures (see figure 1.1). The smallest is the planetary system around a star. Such a system has a size of about 10 light hours. The collection of stars, in turn, make up a galaxy (with about 10^{11} stars per galaxy) having a size of about 20 kpc; the galaxies themselves combine to form groups and clusters with a typical size of a few Mpc. The clusters are part of superclusters with sizes of 50 Mpc or so. And the entire observed region of the universe has a size of about 6000 Mpc.

If you assume that a penny, with a size of about 1 cm, represents the planetary system, the nearest star to us will be about 60 meters away and the size of our galaxy will be 700 kilometers! This should tell you how tiny our entire planetary system is compared to the galaxy we live in. The size of a cluster of galaxies will be 1000 times bigger than this and the entire observed universe is another 1000 times larger!

At the bottom of the ladder, we have the planets which are orbiting the Sun. What about larger structures in the hierarchy? It turns out that none of the structures described above are static. (See table 1.1.) The Sun – or, for that matter, any other star in a galaxy – moves in the galaxy with a typical speed of about 200 km per second (km s^{-1} for short). Even our galaxy is not at rest. It is moving towards the Andromeda galaxy with a speed of about 100 km s^{-1}. Further, the entire Local Group is moving coherently (somewhere towards the Virgo supercluster) with a speed of about 600 km s^{-1}. It is a delicate balance between motion and the gravitational force which keeps our universe going.

Table 1.1 *Structures in the universe*

Structure	Size	Speed
Planetary system	6×10^9 km	30 km s^{-1}
Galaxy	20 kpc	200 km s^{-1}
Local Group	1 Mpc	100 km s^{-1}
Cluster	5 Mpc	1000 km s^{-1}

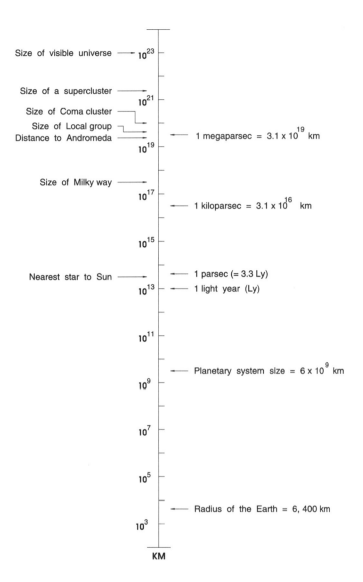

Figure 1.1 Our universe contains a hirearchy of structures from the planets to the superclusters. The smallest unit is the planetary system containing the Sun and the planets which revolve around it. The Sun itself is one among 100 billion stars which make up our Milky Way galaxy. Our galaxy is a part of a Local Group of 30 or so galaxies, all bound together gravitationally and moving under each other's influence. The universe is made of several such groups and clusters of galaxies distributed uniformly on the average. The length scales corresponding to these structures are shown in this figure.

Given the sizes of various systems and the speed with which individual objects move in these systems, one can calculate the typical orbit time for each of these structures. At the smallest scale, the Earth takes one year to go around the Sun, while the farthest planet, Pluto, takes about 250 years. The timescale associated with galaxies is considerably larger. For example, a star like the Sun will take about 200 million years to complete one orbit in our galaxy; and a galaxy will take about 5 billion years to go from one end of a cluster to the other end. It is interesting to compare these numbers with the ages of different objects in our universe. It is estimated that the age of the Earth is about 4.6 billion years; in other words, the Earth has completed more than 4 billion orbits around the Sun since its formation. The typical lifetime of a star like the Sun is about 10 billion years, which is a factor of 50 higher than the orbital time in the galaxy, i.e., a star like the Sun can make about 50 orbits in its lifetime. The ages of bigger structures like galaxies, clusters etc. are far less certain, but – as we shall see in the later chapters – are likely to be about 13 billion years. This is the timescale over which most cosmic phenomena take place.

1.2   The great questions

It is not only as regards dimension that our universe exhibits a fascinating variety. The physical phenomena at different scales also vary tremendously. The stars in our galaxy, for example, come in many different sizes and ages. There are stars in their infancy, middle age and old age; we also see remnants of dead stars in the form of what are called, 'white dwarfs', 'neutron stars' and possibly 'black holes'. There are regions of the galaxy containing large quantities of gas and dust, which behave very differently from the luminous stars. At the next level, galaxies themselves come in tantalizing shapes and sizes and exhibit very different properties; and our universe also contains enigmatic objects like 'quasars' which we have not even mentioned in this chapter.

The description given above raises several key questions: How did all these structures – from planets to superclusters – form? When did they form and what physical processes shaped their properties? How are they distributed in space? What will happen to these structures and the universe in the future?

Modern cosmology has found answers to many of these and related questions. It seems likely that the first galaxies formed nearly 12 billion years ago, and stars and planets formed out of the material in the galaxies. Larger structures, like clusters and superclusters, are likely to have originated from the galaxies coming together under mutual gravitational attraction. All these phenomena seem to be understandable in terms of the basic laws of physics which govern the behaviour of matter and radiation. We shall discuss these issues in detail in this book.

To appreciate and understand these answers, we first need to grasp the way the laws of nature function and affect the behaviour of matter and radiation. This is the task we shall turn to in the next chapter. In chapter 3 we shall briefly describe how astronomers gather information about the universe. Chapter 4 will elaborate on the brief description of the structures in the universe given here, using the concepts developed in chapter 2. The dynamical description of our universe is presented in chapter 5, and the details regarding the formation of various structures are discussed in chapter 6. The next chapter gives an account of the farthest realms of space which astronomy has probed directly, and the last chapter summarizes the current status of our understanding.

2

Matter, radiation and forces

2.1 A microscopic tour

The physical conditions which exist in the centre of a star, or in the space between galaxies, could be quite different from the conditions which we come across in our everyday life. To understand the properties of, say, a star or a galaxy, we need to understand the nature and behaviour of matter under different conditions. That is, we need to know the basic constituents of matter and the laws which govern their behaviour.

Consider a solid piece of ice, with which you are quite familiar in everyday life. Ice, like most other solids, has a certain rigidity of shape. This is because a solid is made of atoms – which are the fundamental units of matter – arranged in a regular manner. Such a regular arrangement of atoms is called a 'crystal lattice', and one may say that most solids have 'crystalline' structure (see figure 2.1). Atoms, of course, are extremely tiny, and they are packed fairly closely in a crystal lattice. Along one centimeter of a solid, there will be about one hundred million atoms in a row. Using the notation introduced in the last chapter, we may say that there are 10^8 atoms along one centimeter of ice. This means that the typical spacing between atoms in a crystal lattice will be about one part in hundred millionth of a centimeter, i.e., about $1/100\,000\,000$ centimeter. This number is usually written 10^{-8} cm. The symbol 10^{-8}, with a minus sign before the 8, stands for one part in 10^8; i.e., one part in $100\,000\,000$. (In this notation – which we shall use frequently – one tenth will be 10^{-1}, one hundredth will be 10^{-2}, etc. To denote two parts in a thousand, we will write 2×10^{-3}. Note that 3×10^3 stands for 3000 while 3×10^{-3} stands for $3/1000$, viz. three parts in a thousand.)

To describe the phenomena at atomic scales, we have to use such small lengths. Just as we introduced new units of length like light year, kiloparsec, etc. to study large astronomical objects, we now need to introduce new units to describe very small lengths. The length 10^{-8} cm, is a convenient unit and is

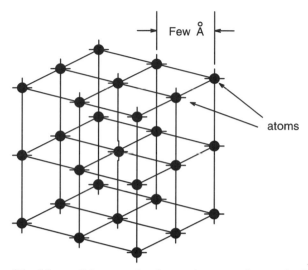

Figure 2.1 Many solids are made of a regular array of atoms in the form of a lattice. The figure shows the 'crystalline structure' of such a simple solid, in which one atom is located at each vertex of a cube. Much more complicated, but periodic, arrangements of atoms are possible in a crystal. The atoms are held in place by electromagnetic forces, and the typical separation between the atoms is a few Angstroms.

called one Angstrom (1 Å, for short). So the separation between the atoms in a crystal is about one Angstrom. A small cube of a solid, 1 cm in the side, will contain about 10^{24} atoms.

What keeps these atoms in such a regular array? It turns out that there are certain forces which act between these atoms which bind them together in the form of the crystal lattice and give the solids their characteristic properties. (We shall say more about these forces in section 2.2.) You know from experience that it takes fair amount of effort with an ice pick to break an ice cube into smaller pieces. In this process, one has overcome the forces that hold the crystal together by supplying energy to the block of ice.

Let us perform a 'thought experiment' by imagining that this process is continued indefinitely. What will happen if we keep supplying more and more energy to a solid body? The answer should be clear from your everyday experience. Heat is a form of energy, and heating a piece of ice is one way of supplying it with more and more energy. And you know that, if we supply sufficient amount of heat energy to a block of ice, it will melt and become water. Heating provides enough energy to break the binding between the atoms in a

crystal lattice, thereby overcoming the rigidity of the ice. The same phenomenon takes place in any other solid. As we heat a solid – and supply more and
more energy – there occurs a moment when the solid melts and forms a liquid.
(Different solids, of course, will require different amounts of heat energy to be
supplied for their melting; it is comparatively easier to melt an ice cube than a
piece of iron.) It takes about 80 calories of energy to melt one gram of ice. This
quantity, the 'calorie' is a unit for measuring the energy. (It is related to the unit
you keep hearing in connection with exercises and diets, all of which have to do
with supplying or burning of energy.) In physics, it is also usual to use several
other units for energy, of which the most frequently encountered one is an
'erg'. One erg is about 2.38×10^{-8} calories; so we may say that it takes
3.4×10^{9} ergs to melt one gram of ice. The rigidity of ice is now lost and
water 'flows' easily.

We can, of course, continue the above experiment by supplying more and
more heat energy to the liquid. At first, the water will just get hotter and hotter.
Soon, another change will occur and the liquid will become a gas. That is, water
will change to steam. You must have noticed that steam is nearly invisible,
compared to water. This is because the atoms forming the steam are a lot more
sparsely distributed compared to the atoms in water. (One cubic centimeter of
steam will have nearly one thousand times fewer atoms than one cubic centimeter of water.) The gaseous phase is also common to any other solid; by
supplying sufficient energy to molten iron we can produce a gaseous vapour of
iron atoms.

In the above sequence we have come across the three most important states of
matter – solid, liquid and gas. Depending on their internal constitution, some
elements (like iron, copper, etc.) are solids at room temperature; some elements
(like mercury, bromine, etc.) are liquids and some others (like oxygen, hydrogen, etc.) are gases. It is, however, possible to convert most of the elements into
any other phase by either heating or cooling. For example, we know that steam
becomes water when 'cooled' below one hundred degrees centigrade. By cooling
to a temperature of -183 degrees centigrade (that is, 183 degrees below zero),
one can make oxygen into a liquid. (For the sake of completeness, it must be
mentioned that some solids have an amorphous, rather than crystalline, nature.
These are exceptions which need not concern us.)

Since the temperature increases as we add more and more heat energy, it is
clear that temperature is a measure of the energy content of the body. It turns
out that the lowest energy state for any matter occurs at around -273 degrees
centigrade, and it is impossible to cool any body to temperatures lower than this

value. (To cool a body, we have to extract the heat energy from the body; which is impossible if it is already in the lowest energy state). In physics, it is often convenient to use a different unit so that the lowest temperature corresponds to zero degrees. This unit is called the 'Kelvin', and is defined so that zero degrees Kelvin (0 K) corresponds to -273 degrees centigrade. To obtain the temperature in Kelvin, we merely have to add the number 273 to the temperature expressed in centigrade. Thus the temperature -183 degrees centigrade (at which oxygen becomes a liquid) corresponds to $-183 + 273 = 90$ Kelvin.

Starting from a solid, and adding more and more energy, we could convert matter to liquid and gas. What happens if we supply still more energy to a gas? To answer this question we need to take a closer look at the constitution of matter. Consider, for example, the simplest of all gases, hydrogen. The individual unit of a hydrogen gas is actually not a hydrogen atom but a 'molecule' of hydrogen which, in turn, is made of two atoms of hydrogen linked together. (Actually, we have been cheating a little in our description of ice, water and steam too, just to keep the discussion simple. An individual unit is again a molecule of water, made of two atoms of hydrogen and one atom of oxygen.) The energy for binding the two hydrogen atoms together, to form the hydrogen molecule, comes from certain electromagnetic forces between the atoms. It is basically the same kind of force which is responsible for the crystalline nature of the solids discussed above. If we manage to supply enough energy to overcome this binding, then the hydrogen molecule will be disrupted and we will get two individual atoms of hydrogen. From various experiments, it is found that one requires about 7.2×10^{-12} ergs of energy to split a hydrogen molecule into two hydrogen atoms. We again see that processes involving individual molecules or atoms require tiny numbers compared to our everyday experience. To measure the energies involved in atomic and molecular processes, it is convenient to introduce yet another unit of energy, called the electron volt (eV). One eV is equivalent to 1.6×10^{-12} ergs; so the 'binding energy' of a hydrogen molecule is about 4.5 eV.

One possible way of supplying this energy would be to collide two hydrogen molecules together with sufficient speed. When we supply heat energy to a gas, the molecules move with more and more speed. On the other hand, we also know that when we supply heat energy to the gas its temperature goes up. Actually, these two descriptions are the same; the temperature of a gas is merely a measure of the average energy with which the molecules move randomly inside the gas. If the hydrogen gas is at room temperature (of about 27 degrees centigrade, which corresponds to $27 + 273 = 300\,\text{K}$), each of the molecules will have, on the average, an energy of about 0.026 eV. If the temperature is a hundred times larger (i.e., 30 000 K), then each molecule will have an energy which is a hundred times larger, i.e, each molecule will have an energy of 2.6 eV. By heating the gas to about $5.2 \times 10^4 = 52\,000$ K, we can supply each of the molecules with an average energy of about 4.5 eV. When those molecules collide, there is a good chance for the hydrogen molecule to split into two individual atoms of hydrogen. Hydrogen gas at temperatures above this will be made of hydrogen atoms and not hydrogen molecules.

What happens if we persist and supply still more energy? We are now supplying the energy directly to the atoms of hydrogen. So, to answer this question, we need to know something about the nature and constituents of an atom.

An atom is made of a 'nucleus' and particles called 'electrons'. We can model the atom as consisting of a nucleus in the centre with electrons orbiting around it. (It must be emphasized that this picture is an over-simplification. Concepts like orbits are not strictly valid at the atomic scales. Nevertheless, this picture is useful for understanding some of the atomic phenomena and we will continue to use it.) The atomic nucleus itself is made of two kinds of particles, called 'protons' and 'neutrons'. Of these particles, protons carry positive electric charge and neutrons have no electrical charge; thus the nucleus, made of neutrons and protons, will have a net positive charge. The electrons which orbit the nucleus carry negative electric charge. An atom will have equal number of protons and electrons, so that it will be electrically neutral. The size of an atom is a few Angstroms (remember that one Angstrom is 10^{-8} cm); the nucleus is much smaller and measures only about 10^{-13} cm. (Incredible though it might seem, most of the space inside an atom is empty – which means that most of the matter is also hollow!)

Even though the electron and proton carry equal and opposite electric charges, the proton is nearly 2000 times heavier than the electron. The neutron is nearly as heavy as the proton. So for all practical purposes the weight of an atom is contained in its neutrons and protons which are in the nucleus.

In the case of hydrogen, the nucleus has no neutrons, and hence the atomic structure of hydrogen is extremely simple. It consists of a proton and an electron with the latter orbiting around the former. Just like the hydrogen molecule, the hydrogen atom also has a binding energy of its own. The electron and proton in the hydrogen atom exert a force on each other which is responsible for the structural stability of the hydrogen atom. This binding has an energy of about 13.6 eV. If we can supply that much energy to the electron in the hydrogen atom, we can destroy the binding and dissociate the hydrogen atom into its constituents, namely the electron and proton.

This can be done by heating the hydrogen gas to a still higher temperature. We saw that at a temperature of 5.2×10^4 K, each particle has an energy of about 4.5 eV. So, by heating the gas to three times the temperature, that is to about 15.6×10^4 K, we can destroy the binding between the electron and proton in the hydrogen and separate them. Actually, the temperatures we have estimated for splitting hydrogen molecule and hydrogen atom are over-estimates. When the gas is at a particular temperature, not all the molecules in it will have the same energy. Most molecules will have an energy related to the temperature of the gas but there will always be a small fraction of molecules with higher than average energy. When we take into account the distribution of energies in a gas, it turns out that we actually need somewhat lower temperatures than estimated above to split the hydrogen molecule or the atom. This fact, however, is not relevant for the key ideas we want to explore.

What if the gas is not made of hydrogen but some other element, say, oxygen? One atom of oxygen is made of eight electrons and a nucleus containing eight neutrons and eight protons. As we said before, the number of electrons in an atom will be the same as the number of protons in the normal state, keeping the atom electrically neutral. (It is just an accident that the oxygen nucleus has an equal number of protons and neutrons; in general, these two will not be in equal number.) The electrons are bound to the nucleus with different amounts of energy. For example, the electrons which are orbiting farthest from the nucleus feel the least amount of force of attraction and hence are bound comparatively weakly. The farthest electron in the oxygen atom has a binding energy of 13.6 eV. (It is yet another coincidence that the binding energy of the outermost electron in an oxygen atom is the same as that of the electron in a hydrogen atom!) In contrast, the innermost electron in oxygen has a binding energy of about 870 eV. As one supplies energy, the outermost electrons are stripped off first, leaving behind an oxygen nucleus surrounded by the remaining electrons.

A new feature enters the picture when we separate the electrons from the nucleus – or 'dissociate an atom' as it is called. Since a normal atom contains an equal number of protons and electrons, it is electrically neutral. Solids, liquids and gases, made of such atoms, are also electrically neutral. When we remove the electron from the hydrogen atom, the hydrogen atom becomes positively charged due to the presence of an unbalanced proton in the nucleus. The electron which is liberated, of course, carries the negative charge. Similarly, if we knock out two electrons, say, from the oxygen atom, then two of the protons in the oxygen nucleus will have unbalanced excess charge, and such an atom will behave as though it carries two units of positive charge. It is usual to call it an oxygen 'ion' (in contrast to an oxygen *atom*, which is neutral). Thus, removing two electrons from a neutral atom leaves behind a positively charged ion carrying two units of positive charge. For this reason, dissociation of an atom is also called 'ionization'. This process of ionization converts matter from an electrically neutral state (like solid, liquid or gas) to an electrically charged state containing positive and negative charges. This fourth state of matter is called 'plasma' and plays a vital role in astrophysics (see table 2.1).

At the atomic level, all matter is made of protons, neutrons and electrons. One can classify all matter by giving the number of protons (which is called the atomic number) and the total number of neutrons and protons in the nucleus (called the atomic weight, since this number tells us how heavy the atom will be).

Each chemical element has a unique atomic number, starting from 1 for hydrogen, 2 for helium, etc. The chemical properties of the element are essentially decided by the number of electrons in the element which, of course, is equal to the number of protons in the nucleus. The atomic weight of an element, however, is not quite unique. Take, for example, hydrogen, which has only one proton in the nucleus and hence has an atomic weight of unity. There exists another element called deuterium which has one proton and one neutron in its nucleus (and one electron orbiting around it). Clearly, both deuterium and hydrogen have the same number of protons (and electrons) and hence the same atomic number; but the atomic weight of deuterium is 2 (because it has a proton *and* a neutron in the nucleus) while that of hydrogen is 1. Since the chemical properties of the elements are determined by the number of electrons, both deuterium and hydrogen have very similar chemical proper-

Table 2.1 *States of matter*

Name	Structure
Solid	rigid; crystalline
Liquid	no rigidity; well defined surface
Gas	randomly moving molecules; no rigidity or shape
Plasma	randomly moving ions and electrons

ties. (For example, two atoms of deuterium can combine with one atom of oxygen and form 'heavy water', denoted by D_2O, just as two atoms of hydrogen can combine with one atom of oxygen and form ordinary water, H_2O.) Similarly, normal helium has an atomic number of 2 (two protons and two electrons) and atomic weight of 4 (two protons and two neutrons in the nucleus). However, there exists another species called helium-3 which has an atomic number of 2 (two protons and two electrons) but an atomic weight of 3 (two protons and *one* neutron in the nucleus). Once again the chemical properties of normal helium (which may be called helium-4) and helium-3 are quite similar. Such species – like hydrogen and deuterium, or helium-3 and helium-4, – are called 'isotopes'. Though the chemical properties of isotopes of the same element will be similar, their atomic nuclei are quite different and behave differently.

As the atomic number increases, the atoms become bigger and, in general, exhibit complex properties. Naturally occuring elements have atomic numbers ranging from 1 to 92. For example, aluminium has an atomic number of 13 and atomic weight of 27. This means that an atom of aluminium contains 13 protons (and, of course, 13 electrons). Further, since the atomic weight is 27 and the atomic number is 13 there must be $27 - 13 = 14$ neutrons in the nucleus (see figure 2.2).

You will, however, be surprised to know that all the elements do not occur with equal abundance in the universe. (See table 2.2.) It turns out that nearly 90 per cent of the matter in the universe is made of hydrogen and 9 per cent is helium. The remaining 100 odd elements, called heavy elements, make up only about 1 per cent of the matter in the universe. (Of course a planet like the Earth has a lot more of the heavy elements; but the Earth is not a typical example for computing cosmic abundances.)

Table 2.2 *The five most abundant nuclei in the universe*

Element	Symbol	Atomic number	Atomic weight	Abundance
Hydrogen	^{1}H	1	1	90%
Helium	^{4}He	2	4	9%
Oxygen	^{16}O	8	16	0·06%
Carbon	^{12}C	6	12	0·03%
Nitrogen	^{14}N	7	14	0·01%

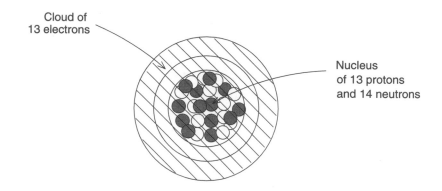

Cloud of
13 electrons

Nucleus
of 13 protons
and 14 neutrons

Aluminium atom

Figure 2.2 Normal atoms are made of a nucleus consisting of protons and neutrons with electrons orbiting around the nucleus. The protons carry positive electric charge and the electrons carry negative electric charge. Normally, atoms have an equal number of protons and electrons and hence are electrically neutral. The properties of the various elements are determined by the number of protons and neutrons in the nucleus. The figure shows (schematically) an aluminium atom, which has 13 protons and 14 neutrons in its nucleus with a cloud of 13 electrons orbiting around it. Note that the figure is not drawn to scale; the size of the nucleus is highly exaggerated.

The structural properties of an atom, like its binding energy, are determined by the electromagnetic interaction between the protons in the nucleus and the electrons which orbit around it. Thus, all physical and chemical properties of gross matter in its four states (solid, liquid, gas and plasma) are governed by the electromagnetic force. The electromagnetic force, however, does not play a major role in the structure of the atomic nucleus itself. As we saw before, the nucleus of aluminium contains 13 protons and 14 neutrons. Since protons are all positively charged particles and like charges repel each other, the nucleus will be blown apart if there is no other force to hold the protons together. Indeed, there exist forces other than electromagnetic force, which are responsible for the stability of the atomic nucleus.

These forces, called the 'strong force' and the 'weak force', have very different characteristics. The strong force is considerably stronger than the electromagnetic force and provides the energy for binding the nucleus together. A typical atomic nucleus has a binding energy which varies from 7 million electron volts to 500 million electron volts. To describe such nuclear processes, we will use the unit 'MeV' which stands for one million electron volts. (We shall say more about these forces in section 2.2.)

The binding energy varies from nucleus to nucleus in a fairly complicated manner. Figure 2.3 shows the amount of energy which is required to pull one nucleon out of the nucleus for different elements. Two features are obvious from this figure. You see that, on the whole, the binding energy per nucleon (i.e. per proton or neutron) increases with atomic number up to iron and then decreases, making iron the most stable of all nuclei. This suggests that elements heavier than iron will behave differently, compared to those lighter than iron, as far as nuclear processes are concerned. In addition to this average behaviour, one can also see significant spikes at some particular elements. For example, helium has an extremely stable nuclear structure, with a fairly high binding energy.

Since the nuclear forces provide the binding for the constituents of the atomic nucleus, we can split the atomic nucleus by overcoming these forces.

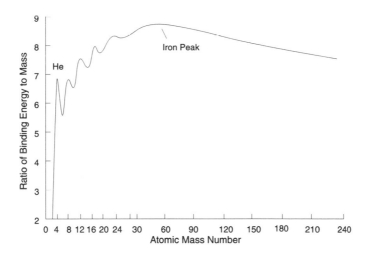

Figure 2.3 Nucleons – which are bound by nuclear forces – can be characterized by a quantity called the binding energy. The higher the binding energy, the tougher it will be to pull a nucleon out of the nucleus. It is clear from the figure that the binding energy (per nucleon) increases up to an atomic number of 56, which corresponds to iron, and then decreases. Among the lighter elements the peak at helium indicates the large amount of energy which can be produced by its synthesis. Elements to the left of iron release energy when they are formed by fusion, while elements to the right of iron release energy when they undergo fission.

The situation here is quite similar to what we did in the case of the hydrogen molecule or hydrogen atom. By colliding two molecules or two atoms with sufficiently high speeds, we could supply enough energy to overcome the molecular or atomic binding. Similarly, by colliding two nuclei together at very high speed we can provide the necessary energy to overcome the nuclear binding. In such – sufficiently energetic – collisions we can actually split the nucleus of an atom. For example, if we hit the nitrogen nucleus with the nucleus of helium, we can obtain two different nuclei; namely, hydrogen and an isotope of oxygen (see figure 2.4). Originally, the nitrogen nucleus contained seven protons and seven neutrons, while the helium nucleus had two protons and two neutrons. In the final state, the isotope of oxygen has nine neutrons and eight protons, while the hydrogen nucleus is made of one proton. Notice that we have now broken new ground. By splitting the nucleus in such a high-energy collision, we have actually converted nitrogen and helium into oxygen and hydrogen. If sufficiently high energies are available, it is indeed possible to convert one element into another.

By processes like this, a nucleus can be split into two; this process is called 'nuclear fission'. It is also possible to fuse two nuclei together and form a new element by a process called 'nuclear fusion'. For example, consider the element

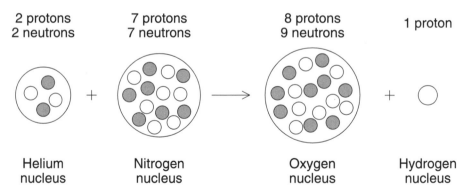

Figure 2.4 The collisions between nuclei can lead to transformation of one element into another. For example, this figure shows the nuclear reaction in which a helium nucleus (with two protons and two neutrons) collides with a nitrogen nucleus (with seven protons and seven neutrons) and forms an isotope of oxygen nucleus (with eight protons and nine neutrons) and a hydrogen nucleus made of a single proton. Such nuclear reactions can absorb or release large quantities of energy, and play a key role in the evolution of stars.

deuterium, which has an atomic nucleus made of one proton and one neutron. By colliding two deuterium nuclei one can produce a nucleus of helium. But since all atomic nuclei are positively charged and like charges repel, the nuclei of two deuterium atoms will have a tendency to repel each other. To fuse these two nuclei together, one has to overcome the repulsion by colliding the nuclei at very high energies. At such high energies, the nuclei can come very close to each other and – at such close distances – the nuclear attraction can overcome the electrostatic repulsion. It follows that nuclear fusion can take place only when the particles are moving with very high speeds, that is, at very high temperatures.

Fission and fusion of nuclei involve changes in energy. As we saw before, the binding energy is different for different elements. Suppose we fuse the nuclei of *A* and *B* and obtain the nucleus of *C*. If the binding energy of *C* is lower than the combined binding energy of *A* and *B*, then this process will release the difference in the energy. On the other hand, if *C* has higher binding energy, then one needs to supply energy to trigger this process. Whether a given process will release or absorb energy depends on the atomic number, with the transition occuring at the atomic number of 56, which corresponds to iron. We saw earlier that iron has the maximum binding energy per nucleon, so that the binding energy curve in figure 2.3 is sloping downwards to the right of iron. In this region we can break up an element to form *lighter* elements which have higher binding energy, i.e., are more stable. Such a process is preferred by nature, and elements heavier than iron – in general – release energy when split up. The situation is reversed for elements which are lighter than iron. Here we can increase the binding energy by fusing two light elements together. Once again such a process is preferred, and energy will be released in the nuclear fusion of lighter elements. In summary, we may say that heavier elements like uranium will release energy when undergoing fission while lighter elements like hydrogen will release energy during fusion. Of course, in order to fuse two nuclei together, we have to overcome the electrostatic repulsion, and hence such a process can take place only at very high temperatures.

The processes described above form the basis of the release of nuclear energy both in nuclear reactors and in nuclear bombs. The hydrogen bomb, for example, is based on the fusion reaction. Since fusing two hydrogen nuclei can release about 0.4 MeV of energy, one gram of hydrogen (containing 6×10^{23} atoms) undergoing fusion will release an energy of about 10^{17} ergs. For comparison, one gram of coal, when burnt, produces only 10^{11} ergs of

19

energy, showing how vastly more powerful nuclear processes are. At a fundamental level, this difference is due to the difference in strengths between
nuclear and electromagnetic forces. The burning of coal is a chemical reaction
governed by electromagnetic forces, while the fusion of hydrogen is governed
by nuclear forces.

2.2   Forces of nature

In the previous section, while discussing the behaviour of matter in different
conditions, we mentioned the kinds of forces which act between the particles,
providing the dynamism behind different phenomena in nature. We will now
take a closer look at these forces and their physical effects.

For convenience of discussion, one can divide the forces of nature into four
categories: electromagnetic force, strong (nuclear) force, weak (nuclear) force
and gravitational force. This division does not mean that all these forces are
actually distinct from each other. We have definite evidence that the electromagnetic and weak nuclear forces are merely two facets of a single entity called
the 'electro-weak' force. It is possible that, at some future date, we will even
have a unified description of all the four forces listed above. But to simplify our
discussion it is better to treat them as distinct.

Electrical forces have been known to mankind for a very long time. The
electric force acts between particles carrying electric charge. Particles can have
positive, negative or zero electric charge. (The last category is called 'neutral'
particles.) A positively charged particle will repel another positively charged
particle while attracting a negatively charged particle; similarly a negatively
charged particle will repel a negatively charged particle but will attract a positively charged particle. Neutral particles do not feel any electrical force. The
force of attraction or repulsion increases in direct proportion to the electric
charge of the particle. That is, a particle carrying twice the amount of electric
charge will exert a force which is twice as large. The force also decreases with
increasing distance between the charged particles; when the distance is
doubled, the force falls by a factor of $2 \times 2 = 4$. If the distance is tripled, it
falls by a factor of $3 \times 3 = 9$ etc. Such a force law is said to satisfy the 'inverse
square' relation.

Just like electric forces, magnetic forces have also been known to humanity
for centuries. What was not known to ancient civilizations was that electric and
magnetic forces are essentially the same. Magnetic force is produced by moving

electric charges and is not a separate kind of force caused by some new kind of 'magnetic charge'. The moving charges constitute what is usually called an electric current, and it is this electric current which produces the magnetic field. Thus one should view the electromagnetic force as a single entity.

These electromagnetic forces also determine the overall structure of matter. The electrons in an atom interact with nuclei through the inverse square force described above. The force between atoms in a crystal lattice is also of electrical nature. Even though individual atoms are electrically neutral, the currents produced by the orbiting electrons inside the atom exert a force on the nearby lattice atom. This force is the cause of the binding energy of a crystal lattice. We saw different manifestations of this force in atomic and molecular systems in the last section.

When we move to the domain of the nucleus, we come across the strong nuclear force, which is very different from the electromagnetic force. To begin with, it is considerably stronger than the electromagnetic force. That is, two protons which are sufficiently close together will feel a nuclear force which is much stronger than their electrostatic repulsion. But the strong force is felt only when two particles are very close to each other. If two protons are separated by a distance of about 10^{-13} cm, then the strong force is 100 times larger than the electrical force. But if the protons are moved apart to 10^{-12} cm, then the strong force decreases by a factor of about 10^5 while the electrical force decreases only by 10^2. So, as the distance increases, strong nuclear forces tend to be ineffective compared to electrical forces (see figure 2.5).

We saw that electrical forces exist only between particles carrying an electric charge and neutral particles are unaffected by them. On the other hand, protons and neutrons feel nuclear forces while electrons do not. One may say that protons and neutrons carry 'strong charge' (just as protons carry 'electric charge') while electrons are 'neutral' as far as strong nuclear forces are concerned. In general, physicists classify elementary particles as 'leptons' and 'hadrons', of which leptons do not feel the strong force while hadrons do. Neutrons and protons are examples of hadrons, while electrons are leptons.

In addition to the strong nuclear force, the elementary particles also feel a weak nuclear force. This force has similarities to both the strong and electromagnetic forces but, at the same time, has some vital differences. To begin with, it is about 10^{11} times weaker than the electromagnetic force. It also has a much shorter range than the electromagnetic force. In fact, its range is comparable to that of the strong force. But, unlike the strong force, both leptons and hadrons are affected by the weak force.

21

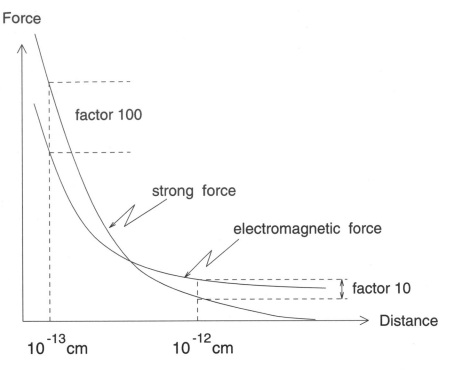

Figure 2.5   This schematic diagram shows the variation with distance of the strength of the electromagnetic and strong force between two protons. The strong force is attractive while the electromagnetic force is repulsive. At very short distances the strong force dominates over the electromagnetic force and the protons can be bound to each other. The strong force, however, decreases rapidly with distance while the electromagnetic force falls only slowly with distance. As a result, at distances larger than about 10^{-12} cm the electromagnetic force will dominate over the strong force.

In reality, the weak nuclear force and the electromagnetic force are only two facets of a single entity called the electro–weak force. When the interacting particles have very high energies (more than about 50 000 MeV), they feel the unified effect of the electro–weak force. But, at lower energies, this force manifests itself as two separate forces.

The last – and, probably, the most important – force of interaction in nature is the gravitational force. When the gravitational force is not too strong it can also be described by an 'inverse square law', just like the electrical forces between two charged particles. There are, however, significant differences between the electromagnetic force and the gravitational force.

Firstly, the gravitational force is always attractive while the electromagnetic force can be either attractive or repulsive depending on the nature of the charges. Secondly, this force is extremely weak; in fact, it is weaker than even the weak nuclear force at comparable distances. The gravitational force between two protons is 10^{36} times weaker than the electromagnetic force between them. Lastly, the gravitational force increases in direct proportion with the mass of the particles. A particle which is twice as massive will exert twice the amount of gravitational force. Since all particles have mass, all particles feel the gravitational force. This is unlike the electromagnetic force, which only affects charged particles and cannot influence neutral particles. The attractive nature of the gravitational force is related to the fact that all particles have positive mass. This is unlike the situation as regards the electric charge, which can be positive or negative.

Comparing the characteristics of the four forces, it is easy to understand why gravity should play the most important role in the case of the large masses encountered in astrophysics, even though it is the weakest of the four forces. The strong and weak forces have an extremely small range and thus are ineffective outside the atomic nucleus. In fact, even in a crystal lattice, the neighbouring atoms interact only through the electromagnetic force; the nuclear forces between the two atoms are completely ignorable. That leaves only the electromagnetic and gravitational forces as contenders for being the most dominant force at large scales. It is indeed true that the electromagnetic force is 10^{36} times stronger than the gravitational force. But it acts only between charged particles. Bulk matter is neutral on the whole because each atom normally has an equal number of electrons and protons. For this reason, massive bodies in astrophysics do not exert electrical forces on each other. This leaves the gravitational force as the only dominant interaction between large, electrically neutral, massive bodies. Most of the dynamics described in the last chapter – planets moving around the Sun, stars moving in a galaxy, galaxies moving in a cluster, etc. – are all governed by the gravitational force.

We mentioned that the gravitational force can be described by an inverse square law when it is weak. It turns out that such a description is only approximate and the true nature of the gravitational force is extremely complex. To be exact, the gravitational force should be properly thought of as 'curvature of space and time'! A simple physical picture of such a description will go somewhat like this: consider a piece of cushion with a perfectly flat surface. If we now place a massive iron ball on this cushion, it will curve the cushion around it. A smaller ball kept in the curved region will roll towards the massive iron ball (see

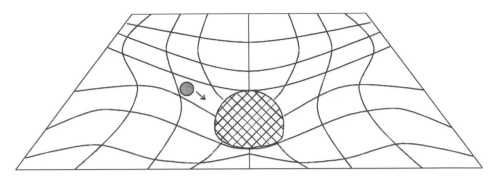

Figure 2.6   According to Einstein's theory, the gravitational force of attraction
between material bodies can be thought of as arising due to the curvature of
space. The idea is illustrated in this figure. The heavy ball kept on a cushion
curves the surface of the cushion. A smaller ball in its vicinity will now roll
down towards it. If we are not aware of the cushion and its curvature, we may
be led to believe that there is a force of attraction between the two balls. The
gravitational attraction between two particles arises in a similar manner due to
the existence of a curved space in their vicinity.

figure 2.6). If we did not know anything about the existence of the curved
cushion, we would have thought that there is force of attraction between the
two iron balls. It turns out that the gravitational force arises due to a similar
'misconception'. Any massive body curves the space around it due to its presence. Any other body located in such a curved space will move in a manner
different from the way it would move in the absence of curvature. If we do not
realize that the space is curved, we will end up thinking that the particles attract
each other by a force. In this description the gravitational force is just a manifestation of the curvature of space and is indistinguishable from it. Incredibly
enough, such a description turns out to be more correct than the elementary
picture based on the inverse square law. The description of gravity as curvature
of space is due to Einstein and is a part of his general theory of relativity.

There is another important modification which the theory of relativity brings
in: energy could be converted into particles with nonzero mass and vice versa,
under suitable circumstances. Hence it follows that *all* forms of energy should
exert a gravitational influence. An electromagnetic field carrying energy, for
example, will produce a gravitational field around it.

2.3 Building blocks of matter

In section 2.1 we explored the nature of matter by supplying to it larger and larger amounts of energy. In this process we first shattered the crystalline structure and converted a solid into a liquid; then we made the liquid into an invisible gas of molecules; by heating the gas – thereby making the molecules collide at larger and larger speeds – we overcame the molecular binding and separated the molecules into constituent atoms; by supplying still more energy, we separated the atom into electrons and ions, thereby converting the matter into a plasma state; finally, by hitting the nuclei with each other at sufficiently high speeds, we found that we can actually split the nucleus into protons and neutrons.

A logical continuation of this 'thought experiment' will be to supply more and more energy to the constituents of the nuclei, viz. protons and neutrons. This can be done by accelerating protons to very high energy and colliding them with, say, other protons. Most of the current research in elementary particle physics uses high energy accelerators which are specifically built for this purpose. A host of new physical phenomena arises in such high energy collisions of particles like protons.

It turns out that we cannot think of particles like protons as truly 'elementary' at very high energies. (At low energies – say, in the study of plasmas at temperatures below 6×10^9 K – one can think of protons and neutrons as stable 'elementary' particles.) If we collide two protons at very high speed, we can produce a host of other particles! This production of new particles is possible because energy can be converted into mass and vice versa. According to the theory of relativity, one gram of matter is equivalent to 10^{21} ergs of energy. Since the proton has a mass of 1.67×10^{-24} gm, its energy equivalent will be about 1.5×10^{-3} ergs, which is the same as 938 million electron volts or, approximately, 1000 MeV. It is convenient to call 1000 MeV 1 giga electron volt, abbreviated as 1 GeV; we may say that the proton has a mass of about 1 GeV using the equivalence between mass and energy. It is an accepted practice in high energy physics to quote all masses in energy units. If we collide two particles with kinetic energy much higher than, say, a few GeV, it is possible to convert part of the kinetic energy into new material particles.

This raises several basic questions. If the collision of two protons can generate new particles, should one imagine that the protons are actually 'made up' of these particles? If so, what will happen when we collide the new set of particles obtained in this process? In other words, do we know what the most fundamental units out of which all matter is made are?

Particle physicists have made significant progress in answering the above questions, even though some fundamental issues are still unresolved. There is considerable evidence to suggest that all matter is made of two basic kinds of particles, called 'quarks' and 'leptons', with no evidence for any further substructure. The forces between these particles are mediated by another set of particles called 'vector bosons'. Let us look at these particles more closely.

The simplest of these three categories is the leptons. There are six different leptons which exist in nature, of which the electron is the most familiar one. There are two more particles called the 'muon' and 'tau' lepton which are identical to the electron, except in their mass. The muon is about 200 times heavier than the electron while the tau lepton is about 3000 times heavier. To each of these three leptons (electron, muon and tau) one can associate a particle called the 'neutrino'. Thus we have three different species of neutrinos, called the electron-neutrino, muon-neutrino and tau-neutrino. These three neutrinos, along with the electron, muon and tau, constitute the six members of the lepton family.

All these six particles – electron, muon, tau and the corresponding neutrinos – exist in nature and have been produced in laboratories during high energy collisions. Of course, only one of these six particles, viz. the electron, exists in the atom under normal circumstances. This does not mean in any way that the other particles are not 'real'. They are all members of the lepton family and enjoy equal status in particle physics.

There are also six different kinds of quarks. They are denoted by the symbols 'u', 'd', 'c', 's', 't' and 'b', which stand for 'up', 'down', 'charm', 'strange', 'top' and 'bottom' quarks. (Some people call the t and b quarks 'truth' and 'beauty'.) Particles like protons, neutrons (and a vast majority of other particles seen in high energy collisions) are made of different combinations of these quarks. For example, a proton is made of two up-quarks and one down-quark, while a neutron is made of two down-quarks and one up-quark. Since all matter has protons and neutrons we might say that – indirectly – all matter contains up-quarks and down-quarks. The other four types of quarks are not present in the normal state of matter. However, they are needed as constituents of several other elementary particles which are produced in high energy collisions. Once again, just like tau-leptons or neutrinos, these quarks are also needed to explain the high energy interactions.

Table 2.3 *Elementary particles*

Particle	Name	Description
Leptons	electron	mass = 0.5 MeV (stable)
	muon	mass = 107.6 MeV (unstable)
	tau	mass = 1784.2 MeV (unstable)
	electron neutrino	
	muon neutrino	mass uncertain
	tau neutrino	
Quarks	up quark	
	down quark	Not seen as free
	top quark	particles; constituents
	bottom quark	of protons and neutrons.
	charmed quark	
	strange quark	
Vector bosons	photon	[electromagnetism]
	W–boson	[weak force]
	Z–boson	[weak force]
	gluons	[strong force]
	[gravitons]	[gravity]

There is, however, one key difference between leptons and quarks. Unlike leptons, the quarks are never seen as free particles in laboratory experiments. The fact that they exist inside a proton can be ascertained only by hitting the proton with very high energy particles which can penetrate into the proton. In such experiments – which probe the proton deeply – one can see that quarks exist as sub-units inside the proton.

The third category of particles is called 'vector bosons'. These particles play the important role of mediating interactions between other particles. According to quantum theory, all forces arise due to the exchange of some fundamental particles. Consider, for example, the electrical repulsion between two electrons. In quantum theory it is described by a process shown schematically in the top left frame of figure 2.7. We assume that the first electron emits a particle (in this case, called a photon) which is absorbed by the second electron. It is this transfer of a photon which actually produces the electrical repulsion between the two particles. In fact, the simplest of the vector bosons is a photon, which is the quanta of electromagnetic radiation. (We shall say more about it in the next section.) All electromagnetic interactions between charged particles can be

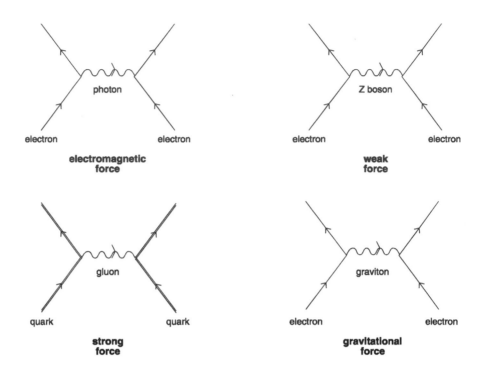

Figure 2.7 According to quantum theory, all forces of interaction are caused by the exchange of certain particles called vector bosons. For example, the electromagnetic force between two charged particles, say electrons, is caused by the exchange of a particle called the photon, which is the simplest of the vector bosons. The strong force between two quarks, say, is caused by the exchange of a vector boson called the gluon. The weak force is similarly caused by the exchange of three vector bosons called the Z-boson, W^+ boson and W^- boson. It is believed that the gravitational force is similarly due to the exchange of a particle called the graviton.

thought of as arising due to the exchange of photons. Other forces of interaction are mediated by some other vector bosons. The weak nuclear force between quarks or leptons is mediated through the exchange of certain particles called 'W' and 'Z' bosons, and the strong nuclear force between the quarks is mediated by a set of eight particles called 'gluons' (see figure 2.7). The gravitational interaction is thought to arise out of certain particles called gravitons. Of all these particles, photons are the most familiar, and we shall discuss them at length in the next section; the W and Z bosons have actually been produced in the laboratory, while the gluons are seen in some of the particle interactions indirectly. Thus the existence of these particles is well founded. The gravitons have not been seen directly and should be thought of as a theoretical speculation at present.

Thus, according to our current understanding, we can interpret all forces of nature using the picture of quarks, leptons and vector bosons. Of these, quarks and leptons act as basic building blocks of matter, while the vector bosons mediate the interaction between the particles.

For the sake of completeness, we must mention another complication. It turns out that every particle in nature has an 'antiparticle'. For example, there exists a particle called the positron which is identical to the electron in all respects except the electric charge. A positron carries positive charge while the electron, as you know, is negatively charged. (The positron should not be confused with the other positively charged particle we have come across, viz., the proton. The proton is nearly 2000 times heavier than the electron. What is more, the proton is made of sub-units called quarks. The positron, on the other hand, has the same mass as the electron and – as far as we know – neither the positron nor the electron has any substructure.) Particle physicists call the positron the 'antiparticle' of the electron. All other elementary particles have their own antiparticles. The proton, for example, has the antiparticle 'antiproton' which is negatively charged. Just as a proton is made of quarks, an antiproton is made of antiquarks. Except for the electrical charge, the antiproton is identical to the proton and, in particular, has the same mass as the proton. (Once again, you should not confuse the antiproton with the electron.)

Though the description of antiparticles might sound like science fiction, it is not. Positrons, antineutrons, antiprotons and many other kinds of antiparticles are routinely produced in the laboratory in the course of high energy interactions. The principle behind the production of any such particle is the same: in a sufficiently high energy collision one can convert part of the kinetic energy into matter; by suitable choice of the colliding particles one can produce the antiparticles mentioned above.

The material world which we see around us is made of protons, neutrons and electrons, i.e., it is made of matter. One can imagine another (hypothetical) world made of antiprotons, antineutrons and positrons as the constituents of (anti)atoms. For reasons which are not very clear, there exists a deep asymmetry between matter and antimatter in the world around us. Simply stated, we see a lot of matter, but virtually no antimatter. We shall see in a later chapter that this fact can have important cosmological implications.

It must also be remembered that the above picture is based on the experimental results which are currently available in the laboratory and on certain theoretical models which are developed using them as inputs. At present, laboratories can only probe phenomena taking place at energy scales less than about 100 GeV. Consequently we are in no position to detect a particle in the laboratory if its mass is considerably higher than 100 GeV. (For comparison, remember that the proton has a mass of about 1 GeV.) The behaviour of particles at higher energies is based on theoretical conjectures. It is often possible to come up with *different* theoretical models which can explain all known experimental observations at low energies. But each of these models will postulate the existence of new kinds of particles which, because of their high mass, would not have been seen in the laboratories yet. For example, there exists a class of theoretical models called 'supersymmetric models' which postulates the existence of several new elementary particles in addition to the quarks, leptons and vector bosons described above. So far, we have not seen any evidence in the laboratory for the existence of these new elementary particles. Very often, cosmological and astrophysical considerations could provide information about such particles – if they exist. We shall see in later chapters that such models may have very important astrophysical and cosmological consequences. Conversely, it will be possible to put constraints on such models based on cosmological and astrophysical observations.

It can also happen that some of the specific properties of the particles are different in different models. Take, for example, the electron–neutrino, usually called just the 'neutrino'. In most models, this particle is assumed to have zero mass. Experimentally, it is known that the mass of the neutrino is smaller than about 12 eV. There are, however, theoretical models in which the electron neutrino can have a mass smaller than this value, say about 10^{-5} eV. It is also possible to produce models in which each of the three neutrinos has a mass of 10 eV, say. Such conjectures, that neutrinos have a tiny mass, will not contradict any of the laboratory results known to us. However, they can have very important cosmological consequences.

2.4 Radiation: packets of light

Radiant energy plays a vital role in astrophysical scenarios, and often appears as an equal partner to matter in determining the nature of physical processes which govern different structures. So we shall next consider different types of radiant energy which exist in nature and how they interact with matter.

You may be familiar with wave-like phenomena in several everyday contexts. If a stone is thrown into a pond, it will produce a concentric wave pattern of ripples on the water surface. This wave will travel outward from the point at which the stone hits the water. With any such wave, one can associate four important parameters (see figure 2.8). The first one is called the 'wavelength', which is the distance between two crests of the wave at any given time. The second quantity is the 'frequency' of the wave, which tells us how many times the wave vibrates in one second at any given location. The third is the 'amplitude' of the wave, which is an indication of how big a disturbance the wave is creating on the surface of the water. Finally, one can also associate a 'speed of propagation' with the wave. This decides how fast the disturbance moves from one point to another along the surface of the water. In the case of water waves, the vibrations consist of movement of water particles on the surface of the pond.

One can, of course, think of many other kinds of waves in nature in which different physical quantities vary periodically in time. For example, sound waves consist of oscillations in the density of air through which the sound is propagating. Normal conversation will produce sound waves with a wavelength of about 35 to 70 cm, corresponding to a frequency of about 500 to 1000 cycles per second.

The wave phenomenon which is of most importance in nature corresponds to the propagation of an electromagnetic disturbance. An 'electromagnetic wave' is a periodic disturbance in the electric and magnetic fields at a given

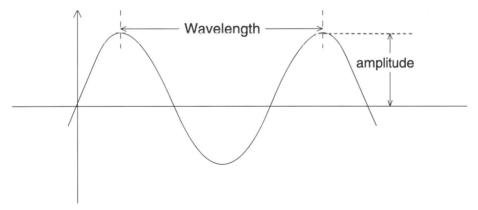

Figure 2.8   With any wave-like disturbance we can associate a wavelength and amplitude. The wavelength is the distance between two crests of the wave, and the amplitude is the height of the wave.

location which is propagating with definite speed. You may wonder why such a wave should have any importance at all. The reason is that normal light is nothing but an electromagnetic wave. Just as with any other wave, we can associate a frequency, wavelength, amplitude and speed of propagation with a light wave. The human eye is sensitive to light waves with wavelengths between 4000 Å and 7000 Å. This wavelength range corresponds to the frequency range from 7.5×10^{14} cycles per second to 4.3×10^{14} cycles per second. One often calls the unit 'cycles per second' Hertz (Hz, for short) for the convenience of writing. All forms of electromagnetic radiation – and, in particular, visible light – can travel at an enormous speed of 3×10^5 km per second.

Most of the light we see – like the light from the Sun – contains a mixture of radiation at different frequencies. For example, it is well known that when sunlight is passed through a prism it is split into the colours of the rainbow. This is because the prism bends the light with smaller frequency more than the light with higher frequency. Red coloured light has a lower frequency (about 4.6×10^{14} Hz) than blue coloured light (about 6.4×10^{14} Hz) and so the prism manages to split the sunlight into its different components.

Electromagnetic radiation can – and does – exist outside the frequency range which is accessible to the human eye. Indeed, several celestial objects emit radiation which cannot be detected by the human eye or ordinary optical telescopes. The radiation with wavelengths outside the visible band is called by different names, like gamma rays, X-rays, ultraviolet, infrared and radio waves; we shall discuss these bands in detail in the next chapter.

In discussing properties like wavelength or frequency of radiation, there is a tacit assumption that the source of radiation and the observer are at rest with respect to each other. In general, this need not be true; for example, we may receive radiation from a star or a galaxy which is moving with respect to us. In those circumstances, an interesting phenomenon called the 'Doppler effect' takes place (see figure 2.9). If the source is emitting radiation at a given wavelength (say, λ_0) and is moving towards us with some speed (say, v), then we will detect the radiation at a shorter wavelength (say, λ_1). Similarly, if the source is moving away from us, we will detect the radiation at a longer wavelength λ_2. Of course the frequency of the wave will increase when the wavelength decreases and vice versa. When the wavelength increases, it is termed 'redshift'; when it decreases, we call it 'blueshift'. (In fact, this is a phenomenon which is familiar to all of us: we hear the whistle of an approaching train at a higher pitch – which is due to the Doppler effect of sound waves.)

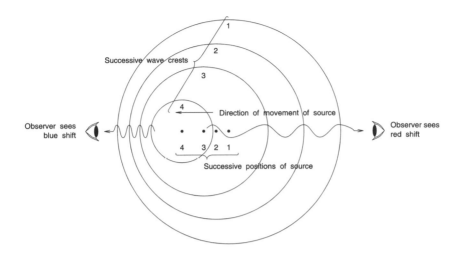

Figure 2.9 The Doppler effect may be thought of as arising due to the compression of the wave crest in the direction of motion of the source. The figure shows four successive crests emitted by a source which is moving to the left. The numbers label the position of the source when it emitted each of the wave crests. The radiation received by the observer at the left is blueshifted, while that received by the observer at the right is redshifted.

The actual value of λ_1 or λ_2 will depend on the speed v of the source. If this speed of the source is much less than the speed of light, then the relative shift in the wavelength will be given by the ratio between the speed of the source and the speed of light. For example, consider a source which is moving at one per cent of the speed of light. Then its wavelength will change by one per cent; if the original wavelength was 5000 Å, it will now change to 5050 Å or 4050 Å depending on whether the source is receding or approaching.

The Doppler effect is an extremely valuable diagnostic tool in astronomy and cosmology. In many situations we will be able to figure out the amount of shift in the wavelength of light emitted by astronomical sources. (We will see specific examples of this in later chapters.) Using the Doppler effect, we can relate the observed shift to the speed of the source. This happens to be the most effective way of obtaining information about the motion of the cosmic sources.

We said that electromagnetic radiation exists in a wide range of wavelengths. Some of the properties of the radiation remain the same throughout this range, while other properties change with the wavelength. For example, all forms of electromagnetic radiation travel with the same speed (viz. at the speed of light) through space. All of them are vibrations in the electromagnetic field; the higher the frequency, the more rapid these vibrations are. There are, however, several other properties of the electromagnetic radiation which vary drastically as the wavelength is changed. In particular, the way radiation interacts with matter depends very much on the wavelength. Consequently, the emission and absorption of radiation by matter depends on the wavelength. You know from experience that visible light cannot pass through human flesh while X-rays can. This is due to the fact that the matter in human flesh interacts differently with X-rays compared to visible light, even though both of them are electromagnetic radiation. We will now look more closely as to how electromagnetic radiation is emitted and absorbed.

2.5   Interaction of matter and light

To understand how radiation is emitted and absorbed by matter, we have to appreciate an intrinsic duality in the description of nature which goes under the name 'the quantum principle'. So far we have talked about electromagnetic radiation as a wave. It turns out that one can think of electromagnetic radiation as made of packets of energy called 'photons'. The energy of an individual photon depends on the frequency of the radiation, and the higher the frequency, the higher the photon energy. An individual photon of wavelength 5000 Å has a frequency of 6×10^{14} Hz and an energy of 4×10^{-12} ergs, which is about 2.5 eV. A photon with twice this frequency will have twice as much energy. One can think of electromagnetic radiation either as a wave or as a bunch of photons. A beam of light can carry more energy either by having more photons of the same energy or by having the same number of photons each of which has more energy.

Incidentally, this duality between wave and particle is a general feature of nature. An electron, which you may think of as a particle, often behaves as though it is a wave with a wavelength determined by its energy. According to quantum theory, even cricket balls and chalk pieces have wave-like features associated with them. However, the wavelengths associated with such macroscopic objects are so tiny that – for all practical purposes – we can treat the wavelength as zero and ignore the wave nature of such an object. More generally, the wave nature will make its presence felt when the wavelength (which is related to the energy) becomes comparable to other length scales like, for example, the size of the measuring apparatus, and the particle nature will be important when the wavelength is small compared to other length scales. Because of this reason, long wavelength electromagnetic radiation – like radio waves, infrared or even visible light – can be thought of as a wave; at the other extreme, short wavelength radiation – like X-rays or gamma rays – should be considered as being made up of photons.

At the tiny scales of atomic systems, we have to take into account the particle nature of light. In fact, atomic systems emit and absorb radiation in the form of photons. Consider, for example, the electron orbiting the proton in a hydrogen atom. It turns out that such an electron cannot have any arbitrary amount of energy while in the hydrogen atom. The lowest energy it can have, when it is most tightly bound to the proton, is called the 'ground state energy' and corresponds to -13.6 eV. (It is a common practice to add a minus sign to the energy in order to show that the electron is bound inside the atom. By convention, a free electron at rest has zero energy and a free, moving electron has positive energy. We know that we have to supply 13.6 eV of energy to 'free' the electron. So the electron in a hydrogen atom must have *less* energy than the free electron – which, by convention, has zero energy; obviously, the energy of the electron bound in an atom should be indicated by a negative number). In addition to the ground state, the electron can also exist at several higher energy levels called 'excited states'. The first excited state has an energy of $-[13.6/(2 \times 2)]$ eV $= -3.4$ eV and the second excited state has an energy $-[13.6/(3 \times 3)]$ eV $= -1.5$ eV etc. Note that -3.4 is *less* negative than -13.6; so -3.4 is larger than -13.6. Thus an electron with energy -3.4 eV has higher energy than an electron with energy -13.6 eV.

It is not possible for an electron in the hydrogen atom to have energies between, say, -13.6 eV and -3.4 eV or between -3.4 eV and -1.5 eV! We say that the energy of the electron is 'quantized'; it can take only specific values. Equivalently, we may say that the electron can exist in only certain definite energy levels. This is one of the key predictions of quantum theory, which is, of course, confirmed in laboratory experiments.

It is possible for an electron to jump from a higher energy level to a lower energy level on its own, changing its energy. When an electron jumps from the first excited state (which has an energy of -3.4 eV) to the ground state (with an energy of -13.6 eV), its energy changes by the amount 10.2 eV, which is the difference in these energies. This energy is emitted as a photon with a frequency of 2.5×10^{15} Hz and a wavelength of about 1216 Å. (Remember that there is a definite relation between the frequency and energy of a photon; so, given the energy, we can determine the frequency.) This is an example of the mechanism by which a hydrogen atom can emit electromagnetic radiation. Similarly, by making transitions between different energy levels, the atom can emit radiation of different wavelengths (see figure 2.10).

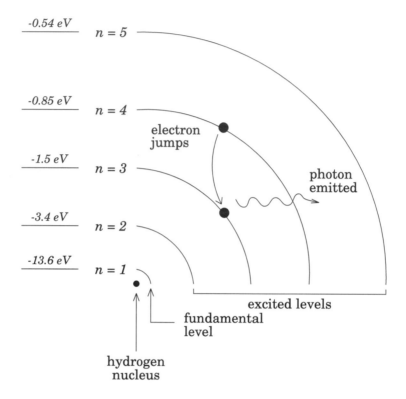

Figure 2.10   An electron orbiting the nucleus can exist only with certain definite values for its energy. The lowest energy level for an electron in a hydrogen atom has an energy of −13.6 eV. (This is the level marked $n = 1$ in the figure and corresponds to the electron being closest to the nucleus.) The minus sign is an indication of the fact that the electron is 'bound' to the nucleus; in other words, it takes 13.6 eV of energy to free the electron from the nucleus. The next energy level marked, $n = 2$, has an energy of −3.4 eV. This is a higher energy level because it is less negative. The figure shows the first five energy levels in the hydrogen atom. When an electron 'jumps' from a higher level to a lower level, it emits the difference in the energy in the form of a photon. This photon will have a wavelength characteristic of the energy difference between the two levels.

The same physical process is responsible for the emission of radiation in almost all cases. Different atoms will have different sets of energy levels, and the transitions of electrons between these levels will lead to the emission of radiation at different frequencies. In fact, the process is not restricted to atoms. Molecules, for example, have their own energy levels (in addition to the energy levels of the atoms which make up the molecule) and hence can emit radiation by making a transition between two such levels.

The pattern of radiation emitted by a particular atom (or molecule) is called its 'spectrum' or, more precisely, 'emission spectrum'. Since the actual spacing between the energy levels differs from atom to atom (and molecule to molecule) the spectrum of hydrogen will be quite different from, say, the spectrum of carbon dioxide. By studying the spectrum of some source of radiation, one can often pin-point the elements which are present in that source. What is more, by measuring the amount of energy that is emitted in different transitions, one can also determine the relative abundance of elements which are present. In this way the spectrum serves like a signature of a particular molecule or element, and is a very important diagnostic tool in physics, chemistry and astronomy.

A somewhat more exotic version of the above mechanism is responsible for radiation at a wavelength of 21 cm, which is used extensively in astronomy. A wavelength of 21 cm is quite macroscopic and is in the radio band; correspondingly the energy of a photon with this wavelength is extremely tiny and is about 6×10^{-6} eV. Surprising though it might seem, this radiation is emitted by the hydrogen atom! We have seen earlier that the energy levels of electrons in the hydrogen atom are of the order of eV, and you may wonder how radiation with a wavelength of 21 cm could come out of the hydrogen atom. It turns out that particles like the electron, proton, etc. not only carry charge but also have another intrinsic property called the 'spin'. The spin has the effect of making the electron and proton behave like very tiny bar magnets. The force of interaction between the electron and proton is not only due to the electrical attraction but also because of the tiny magnetic force arising from the spins. The magnetic force is very small compared to the electrical force; but this tiny force does contribute to the energy of the electron in the hydrogen atom. The magnetic contribution to the energy depends on whether the spin of the electron and that of the proton are aligned in the same direction or opposite to each other. These two states differ in energy by 6×10^{-6} eV! The electron can make a transition from one of these states to another, emitting the excess energy in the form of a photon. This is how the neutral hydrogen can emit a photon with a wavelength of 21 cm. We shall see, in later chapters, that the 21 cm radiation turns out to be a boon to astronomers.

There is another important piece of information which the spectrum can provide. We mentioned earlier that the wavelength of radiation (as detected by us) could change if the source is moving towards or away from us. Suppose we have a source of gaseous hydrogen emitting radiation at two different wavelengths, say, λ_1 and λ_2. For example, the transition of electrons from the first excited level to the ground state will lead to the emission of photons with wavelength $\lambda_1 = 1216$ Å; the transition from the second excited level to the ground state will produce photons of wavelength $\lambda_2 = 1025$ Å. If the source is not moving, then we will detect photons with these two wavelengths from the source. But if the source is moving towards us, both λ_1 and λ_2 will get shifted to shorter values λ_1' and λ_2'. The percentage of shift in each case will be v/c, where v is the speed of the source and c is the speed of light. Suppose the source is moving towards us with a speed of 300 km s^{-1}, which is one thousandth of the speed of light. Then the wavelengths will change by one part in a thousand. The wavelength $\lambda_1 = 1216$ Å will change to $\lambda_1' = 1215$ Å; similarly $\lambda_2 = 1025$ Å will change to $\lambda_2' = 1024$ Å. But note that such a shift cannot change the *ratio* between the two wavelengths. Originally the ratio between the wavelengths was $\lambda_1/\lambda_2 = 1.186$; though the Doppler effect changes the individual wavelengths to λ_1' and λ_2', it does not change this ratio, and indeed $\lambda_1'/\lambda_2' = 1.186$. This ratio of the wavelengths is known from theory. An astronomer studying the spectrum of radiation emitted by a distant source will first measure such ratios. Comparing the measured ratios with those known from theory, he will be able to conclude that the radiation is emitted, say, from a gas cloud of hydrogen. Once we know that the source is hydrogen, we can compare the observed wavelength λ_1' with the theoretically expected wavelength λ_1 and thus find the speed of the source.

In reality astronomers study not just two lines but a whole sequence of lines from a source. One would typically find that all the lines emitted by the source are shifted by a constant factor. By comparing such a spectrum with a standard spectrum, one can accurately measure the shift in the wavelength and thus obtain the speed of the source (see figure 2.11).

The absorption of radiation proceeds in a similar manner but in the opposite direction. Consider, for example, an electron in the ground state of a hydrogen atom. It cannot jump to a higher energy state on its own since this process would require some energy. But if radiation is shining on the atom, it can absorb a photon of wavelength 1216 Å and make a transition from the ground

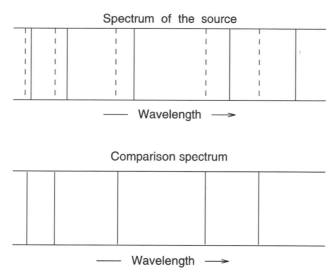

Figure 2.11 The Doppler effect shifts the lines in a spectrum of a source depending on its motion. The top part of the figure shows the actual spectrum of a source. The unbroken black lines correspond to the wavelengths at which radiation was detected. The bottom figure is the comparison spectrum of some element in the laboratory. One notices that each of the lines in the spectrum of the source is shifted systematically to a longer wavelength as compared to the lines in the comparison spectrum. However, the ratio between the wavelengths of any two lines is the same in the two figures. Because of this feature, we can identify the spectrum of the source as due to the element in the bottom figure and attribute the shifts in the wavelengths to the motion of the source.

state to the first excited state; or it can absorb a photon of wavelength 1025 Å and make a transition to the second excited state and so on (see figure 2.12). Other atoms and molecules with different sets of energy levels behave in a similar manner; all of them can absorb photons and make transitions from lower energy levels to higher energy levels.

Just as in the case of emission, an atom (or molecule) can only absorb photons of definite wavelengths when it makes a transition between two energy levels. Suppose we shine light of all wavelengths on a system made of a large number of hydrogen atoms. Photons of most wavelengths will pass through the gas, while photons of specific wavelengths (like 1216 Å and 1025 Å) will be absorbed by the atoms (see figure 2.13). On observing the radiation which has passed through the hydrogen gas, one will clearly see the tell-tale absence of photons at particular wavelengths. Such a spectrum is called an absorption

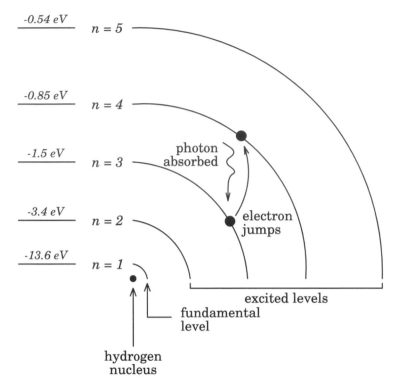

-0.54 eV — $n = 5$

-0.85 eV — $n = 4$

-1.5 eV — $n = 3$

-3.4 eV — $n = 2$

-13.6 eV — $n = 1$

photon absorbed

electron jumps

excited levels

fundamental level

hydrogen nucleus

Figure 2.12 This figure shows the absorption of a photon, just as figure 2.10 showed the emission. The electron can absorb a photon and jump from a lower level to a higher level provided the photon has exactly the energy corresponding to the energy difference between the two levels. If the energy of the photon is different, then it will not be absorbed by the hydrogen atom.

spectrum. By studying an absorption spectrum we can determine the composition of elements in regions through which radiation has been passing.

We described before that the electron in the hydrogen atom is bound to the proton with an energy of −13.6 eV. What happens if we shine photons with higher energy, say 15 eV, on the hydrogen atom? To answer this question we must remember that a free electron which is not moving has zero energy and a free electron moving with some speed has positive energy. If we shine photons with energy 13.6 eV on a hydrogen atom, then the electron can absorb one such photon and reach the zero energy level. This means that the electron has just become free. If we shine the atom with photons of 15 eV energy, then the electron which absorbs the photon will

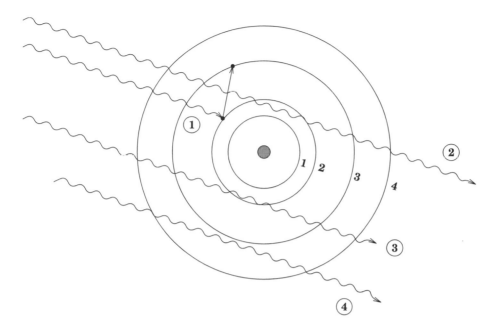

Figure 2.13 When photons of different wavelengths pass through a hydrogen atom, only those with specific wavelength will get absorbed. This is because only photons with certain wavelengths will have energies corresponding to the energy difference between the levels of the hydrogen atom. In the figure the photon marked 1 has been absorbed by an electron which makes a transition to a higher level. Other photons like 2, 3 and 4 pass through the hydrogen atom unhindered. (It is assumed that all the photons 1 to 4 have energy less than 13.6 eV.)

end up with a positive energy of $(15 - 13.6)$ eV = 1.4 eV. That is, the electron will escape from the hydrogen atom and will be moving freely with a kinetic energy of 1.4 eV. We have thus separated the electron from the proton in the hydrogen atom and converted matter from the gaseous state to the plasma state. This process of ionizing the atom (that is, dissociating the atom into an electron and a positively charged ion) is called 'photo-ionization'. (Earlier we discussed a procedure for ionizing atoms by colliding them together; that process is called 'collisional ionization'.) The same principle also works in the case of other systems like heavier atoms, molecules, etc. All of them can be ionized by sufficiently high energy photons. This is one of the reasons why high frequency radiation can cause structural damage to materials. (Incidentally, this discussion also shows that photons with *any* energy greater than 13.6 eV will be absorbed by the hydrogen gas.)

The nuclei of the atoms can also exist in different energy levels. When the nucleus makes a transition between two such levels, it can emit very high energy photons. Since nuclear energy levels run to millions of electron volts, the photons emitted from the nucleus have frequencies upwards of 2×10^{20} Hz; that is, they are gamma rays.

The above description of absorption and emission of radiation is based on the picture of photons. Since radiation can be thought of either as photons or as waves, you may wonder how the same phenomena can be described in terms of waves. For example, are radio waves made of photons? At a basic level the answer is, of course, 'yes'. But as we said before, if the wavelength of radiation is comparable to other length scales involved in the problem, then it is better to think of electromagnetic radiation as a wave. For visible light, the wavelength is like a 50 millionth part of a centimeter; clearly this is tiny compared to everyday scales. Radio waves, in contrast, could be tens or hundreds of meters in wavelength, which could be comparable to or bigger than other length scales in the system. In that case one may think of electromagnetic radiation as 'waves' which are being emitted by the motion of electrically charged particles. Whenever a charged particle moves with a variable speed, its kinetic energy changes, and the difference in kinetic energy could be radiated in the form of electromagnetic radiation. You would have noticed that in any process involving the emission of electromagnetic radiation the energy of the charged particle changes. In the photon picture, this arises because the system emitting the radiation makes a transition from one energy level to another. In the wave picture, this arises because the kinetic energy of the charged particle changes.

One important example of radiation arising from the motion of charged particles is called 'synchrotron radiation'. This radiation arises when a charged particle moves very rapidly in a magnetic field. The key effect of the magnetic field on a charged particle is to make it move in a spiraling manner along the direction of the magnetic field. During such a motion, the particle is constantly accelerated and will radiate energy. Astrophysical systems invariably contain magnetic fields which could influence the motion of the charged particles. As a result, synchrotron radiation is of primary importance in astrophysics and usually occurs in a wave band stretching from centimeters to meters. This is the wavelength below standard AM (around 300 meters wavelength) and FM (around 3 meters wavelength) used in radio transmissions. Radio astronomers have very effectively used synchrotron radiation as a probe of astrophysical systems.

There is one special kind of radiation which plays a vital role in astrophysics, called thermal radiation. This is the radiation emitted by any body due to its temperature. You know that when an iron rod is heated to a high temperature it emits light of predominantly red colour; we say that the iron is 'red hot'. This is merely a particular case of a very general phenomenon. When any material object is kept at some nonzero temperature, it emits radiation of a particular characteristic which depends only on temperature. The radiation will contain a wide band of wavelengths, but will be peaked at a wavelength which decreases with the temperature. Bodies at room temperature (which corresponds to 300 K) will emit most of the radiation in the infrared band. If the temperature is increased to about 10^4 K, the radiation will be mostly in the visible range, and if we heat the system to about 10^8 K there will be copious production of X-rays. The amount of radiation emitted at different frequencies by a body at some temperature is called the 'thermal spectrum' or 'Planck spectrum' (see figure 2.14). At a microscopic level, the Planck spectrum arises because transitions between a large number of energy levels cause emission of radiation in a wide range of frequencies.

In summary, we have seen three different processes for the emission of electromagnetic radiation. The first is due to the transition between different energy levels in a system, like an atom, molecule or nucleus. This produces a spectrum of radiation which is completely characteristic of the source that emits the radiation. Secondly, charged particles moving with changing speed can emit electromagnetic radiation. This process is primarily responsible for the emission of long wavelength radiation like radio waves. Finally, any object at a finite temperature will emit radiation with a very characteristic spectrum called the Planck spectrum. Most of the radiation will be at a wavelength which depends only on the temperature of the body. As the temperature is increased, most of the radiation will be emitted at higher and higher frequencies or – equivalently – at lower and lower wavelengths.

It turns out that there is yet another way of generating photons, which arises from the fact that matter and energy are interconvertible. We saw earlier that there exists an antiparticle for every elementary particle. When a charged particle and its antiparticle collide with each other, their total energy can be converted into radiation. For example, when an electron and positron collide, they annihilate each other, and their total energy is emitted as radiation in the form of two photons. The mass of the electron (or positron) is equivalent to an energy of 0.5 MeV. A photon with this energy will have a frequency of 10^{20} Hz and thus will be in the gamma ray band.

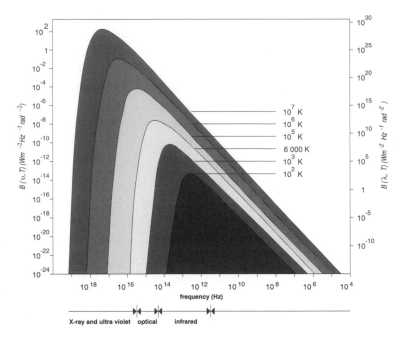

Figure 2.14   Any object kept at some temperature emits a characteristic spectrum of radiation called the Planck spectrum. As the temperature is increased, most of the radiation will be emitted at shorter and shorter wavelengths. The figure shows the nature of emission for bodies kept at temperatures ranging from 100 K to 10^7 K. Bodies at room temperature (300 K) emit most of the radiation in the infrared. Bodies at higher temperatures, say, between 5000 K to 7000 K (as in stars) emit a large amount of visible radiation.

It is also possible to reverse the above processes. Just as a particle and antiparticle can annihilate each other and produce photons, we can also produce particle–antiparticle pairs out of sufficiently energetic photons. For example, if a physical system has large number of gamma ray photons with energies greater than 0.5 MeV, every once in a while a pair of photons will convert themselves into electron–positron pairs. Of course, electron–positron pairs will also – in turn – annihilate and reproduce the photons. These two processes are governed by different factors and will proceed at different rates. A steady state will be reached when the rate of annihilation of electron–positron pairs is balanced by the rate of production of electron–positron pairs from high energy photons. In such a steady state we will have a certain number of high energy photons and a certain number of electron–positron pairs.

The above discussion brings out one feature of high energy interactions which we did not cover in the previous sections. When we discussed the heating of a gas to high temperatures, we mentioned that the system will reach a plasma state made of charged particles like ions and electrons. Such a system will also have a very high temperature. But you will recall from the above discussion that, at any given temperature, there will always be some radiation present which we called the thermal radiation. At high temperatures this thermal radiation will be peaked at high frequencies and will contain a large number of photons with energies greater than, say, 0.5 MeV. But then, such high energy photons are capable of generating electron–positron pairs. So we conclude that any plasma at sufficiently high temperature will have not only ions and electrons (which arise from the dissociation of atoms) but also some electron–positron pairs arising from high energy photons.

What happens when we increase the temperature still further? Soon the thermal radiation will have sufficiently energetic photons to produce copious amounts of other elementary particles like protons and antiprotons. For this reason, the study of very high temperature plasma systems is inextricably linked with the study of high energy particle interactions. We shall see later that such considerations play a vital role in the physics of the early universe.

3

Observing the universe

3.1   The cosmic rainbow

In chapter 1, we did a rapid survey of the universe, listing its contents, and in chapter 4, we plan to discuss these objects in more detail. You may wonder how such a detailed picture about the universe has been put together. This has been possible because we can now observe the universe in a wide variety of wavebands of the electromagnetic spectrum, and virtually every cosmic object emits radiation in one band or another. In this brief chapter, we shall have a rapid overview of how these observations are made. While describing the observational techniques, we will also mention briefly the astronomical objects which are relevant to these observations. These objects are described in detail in the next chapter, and you could refer back to this cha46
pter after reading chapter 4.

It is rather difficult to ascertain when the first astronomical observation was made. Right from the days of pre-history, human beings have been wondering about the heavens and making note of the phenomena in the skies. The earliest observations, needless to say, were made with the naked eye. With the advent of the optical telescope, one could probe the sky much better and detect objects which were too faint to be seen with the naked eye. As the telescopes improved, the quality of these observations increased.

There is, however, an inherent limitation in these early observations. All these observations were based on visible light. We now know that visible light is an electromagnetic wave whose wavelength is in a particular range. The human eye can detect electromagnetic radiation with wavelengths between 4000 Å and 7000 Å. Radiation emitted by astronomical bodies outside this range cannot be detected by normal optical telescopes. Thus, techniques based on optical astronomy, even with the best telescopes, can probe only a tiny band of wavelengths.

Observational astronomy in the wave-bands outside the visible band developed only comparatively recently. This delay was mainly due to two technological problems. To study the universe outside the visible band one has to construct special 'eyes' which will receive radiation and – if necessary – focus them to a point. (This is the task traditionally done by the lens or the mirror in an optical telescope.) Secondly, it should be noted that the Earth's atmosphere absorbs most of the non-optical radiation (see figure 3.2). Hence, to study this radiation, suitable detectors have to be flown above the atmosphere in balloons, airplanes, rockets or spacecraft. The technology to achieve these effectively came into being only in the last sixty years or so. As a result, our understanding of the radiation from faraway sources has increased significantly only in the last few decades. We can now probe wavelengths which are less than one thousand millionth shorter than visible light, or wavelengths which are hundred million times longer than visible light. Most of the information about the universe described in the next chapter comes from probing the universe in this vast range of wave-bands.

As we saw in chapter 2, the radiation at wavelengths shorter and longer than visible light is called by different names: gamma rays, X-rays, ultraviolet, infrared and radio waves (see figure 3.1). Gamma rays are the ones with wavelengths less than 0.1 Å. (This is about one tenth of a size of an atom!) As the wavelength of the radiation decreases, its frequency increases and so does the energy carried by the quanta of radiation. Gamma ray quanta have very high energies and come from the most energetic astronomical sources: from compact pulsars or from distant quasars. (See chapters 4 and 7 for a detailed description of these astronomical objects.) They can also arise from nuclear reactions triggered by the collision between very high energy cosmic particles and gaseous atoms in space.

Slightly higher in wavelength are the X-rays, with a range between 100 and 0.1 Å. These are the same kind of rays which are so frequently used in the medical profession to probe the human body. The cosmic X-rays come essentially from super-heated gas in clusters of galaxies or from streams of gas in binary star systems. The temperature of such gaseous systems can be over a million degrees, and the study of X-rays from clusters of galaxies provides a useful probe of these objects.

Next in the scale is the ultraviolet radiation, with wavelengths from about 100 to 4000 Angstroms. Our Sun is a prominent source of this UV radiation, which causes the tanning of the skin on exposure to sunlight. This radiation is absorbed by the ozone gas in the atmosphere, and UV observations have to be

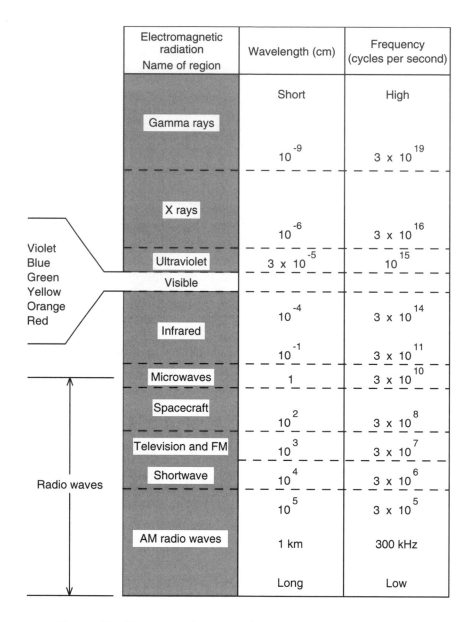

Electromagnetic radiation Name of region	Wavelength (cm)	Frequency (cycles per second)
	Short	High
Gamma rays		
	10^{-9}	3×10^{19}
X rays		
	10^{-6}	3×10^{16}
Ultraviolet	3×10^{-5}	10^{15}
Visible		
	10^{-4}	3×10^{14}
Infrared		
	10^{-1}	3×10^{11}
Microwaves	1	3×10^{10}
Spacecraft		
	10^{2}	3×10^{8}
Television and FM	10^{3}	3×10^{7}
Shortwave	10^{4}	3×10^{6}
	10^{5}	3×10^{5}
AM radio waves	1 km	300 kHz
	Long	Low

Violet
Blue
Green
Yellow
Orange
Red

Radio waves

Figure 3.1   Electromagnetic waves exist in nature with a wide range of wavelengths (and corresponding frequencies). As the wavelength decreases, both the frequency and energy content of the wave increase. The highest energy radiation is called gamma rays, and radio waves are one of the lowest energy waves used by mankind. Visible light, containing the familiar colours, forms a very narrow band of wavelengths.

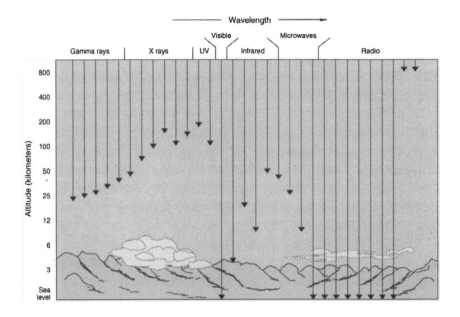

Figure 3.2 Electromagnetic radiation of different wavelengths is absorbed by different amounts in the atmosphere. It is clear from this figure that only radio and visible bands are accessible from the ground. Some of the other bands (like the near infrared) can be studied using balloons, while most others require satellites.

carried out from outside the atmosphere. The major sources of UV emission are the hottest stars and supernova remnants in our galaxy. As the UV radiation passes through space, it is imprinted with information about the tenuous gases which exist in the interstellar medium, and helps astronomers to study these systems.

On the other side of the visible spectrum is the radiation which is slightly longer in wavelength than visible light, called infrared. This is electromagnetic radiation with wavelengths between 7000 Angstroms and about 1 millimeter. We are constantly surrounded by this radiation, which is emitted by objects even at room temperature, and think of it as heat radiation. (The higher the temperature of an object, the shorter is the wavelength at which most of the energy is emitted.) By astronomical standards, infrared radiation is emitted by objects which are rather cool. Most of the cool stars or cold clouds of gas and dust emit in this wave-band. Infrared astronomers can thus actually 'see' these clouds of dust in space, which are invisible at other wavelengths. These gas clouds are the breeding grounds for new stars, and star births are primarily visible in the infrared.

Beyond the infrared lies the vast band of radio waves, which is radiation with wavelengths greater than about 1 millimeter. Radio astronomers regularly scan the sky at wavelengths of a few millimeters to centimeters. Some of the more powerful telescopes can observe the universe at much longer wavelengths. Astronomical sources of radio waves are usually charged particles swirling in cosmic magnetic fields, or neutral hydrogen gas, which is present in vast amounts all over the universe.

Each of these wavelength bands requires special techniques of detection and contains different varieties of information. By putting these pieces of information together one can understand the rainbow of the universe in its full glory. We shall see examples of these in the next few sections.

3.2   Seeing is believing

The oldest and most traditional form of astronomy is optical astronomy, which gathers information about our universe using the narrow band of visible light. Historically, this kind of astronomy developed first for the simple reason that telescopes helped the astronomers to see better.

With the naked eye, one can see a few thousand stars in the sky. The information so gathered is not of much use in understanding the structure or dynamics of celestial objects. There are basically two problems. First, the eye simply cannot resolve fine details in the objects in the sky. The full Moon, for example, shows certain blackish structures in it, but one needs a pair of binoculars or a telescope to resolve them into craters and mountains. By using larger magnification, one can see these details more distinctly. It may seem that by increasing the magnification of the telescope you can eventually see the footprints of Apollo astronauts left on the moon. This, however, is not true because of two other problems. Firstly, there is a limit to the sharpness of the image produced by any telescope, called the 'diffraction limit'. This limit is decided essentially by the size of the telescope's lens. If you magnify too much you don't see finer details but merely make a fuzzy image fuzzier. The solution lies not in increasing the magnification but in going for larger aperture – that is, a bigger lens. A bigger lens also gathers more light and thus helps you to see fainter objects. Secondly, the turbulent motion of air in the Earth's atmosphere also limits the clarity of the image. For large telescopes, this can be the main source of limitation.

The story of optical astronomy was essentially the story of bigger and better telescopes. Historically, the first telescopes which were constructed were based on lenses. One lens (called the 'objective') will act as the front end collecting light and forming an image which is viewed through a second lens (or a combination of lenses) called the 'eyepiece'. Galileo was the first person to investigate the sky systematically with his telescope, which had a lens 4 cm across. With this telescope he discovered the satellites of Jupiter and the phases of Venus. Soon people started building bigger and bigger 'refractors' (which is the technical term for telescopes based on lenses) and one of the largest to be built was a 1 meter instrument at the Yerkes observatory, Wisconsin, completed in 1897. There is, however, a limit to the size of a refracting telescope. A lens can only be supported in its place by structures sticking to its edge. (Otherwise the structure will block the light). This does not work for a very large lens because it will sag under its own weight, affecting the quality of images. The Yerkes lens weighs almost a quarter of a ton and this is just about the limit.

There is, however, another way of making telescopes, which has also been known for centuries. One can use a curved mirror at the back of the telescope to focus light to a point in front of it. One can either build a small cabin at the focus from which observations can be carried out, or one can use a secondary mirror to redirect the light to a more convenient point. This arrangement was first used by Sir Isaac Newton in 1672. The reflecting telescope has a major advantage over the refracting type: a mirror can be supported over the whole of the back surface and hence will sag very little under its own weight. In fact the Prussian astronomer William Herschel built a reflecting telescope in 1789 which was larger than the Yerkes refractor built almost a century later.

The first of the great modern reflectors was the 2.5 meter telescope at Mount Wilson, California, completed in 1917. This was followed by a 5 meter telescope at Mount Palomar, California. Since then telescopes have come a long way; many of the telescopes now in operation or construction are 8 to 10 meters in size. The Keck telescope at Hawaii (10 meter), Magellan at Las Campanas observatory in Chile (6.5 meter), Gemini at Mauna Kea, Hawaii (8 meter) and Subaru at Mauna Kea, Hawaii (8.2 meter) are some of the telescopic giants.

Most of these telescopes have been built on mountains or peaks far from civilizations and high above sea level. This is necessary for better 'seeing': at low altitudes atmospheric twinkling as well as the haze from city lights swamp the faint signals the astronomer is looking for. The best observatories today are located in the wild heights of the Andes, in Chilean mountains or even on top of a huge extinct volcano (Mauna Kea in Hawaii). Still better, of course, would be to put a telescope in a satellite so as to be completely immune to the vagaries of the atmosphere. This hope was realized when the Hubble Space Telescope was launched in 1990. After the initial hiccup due to what the press called the 'billion dollar blunder' and the much publicized, face saving repair mission, the Hubble Space Telescope is now routinely producing very good quality data in optical astronomy.

What do you do with the telescope? Galileo and Newton painstakingly looked through them, recording the observations manually. But the human eye is not a good light detector, and it is a waste to put it at the working end of an expensive telescope. (The eye is not very sensitive to faint objects, it cannot store images for future use or measure brightness or position of stars with any precision!) All modern telescopes send their light signals to a photographic plate or to an electronic detector. Photographic plates with sophisticated processing techniques and colour filters have given some of the most stunning pictures of the universe in recent years. Filters also allow the astronomer to study the stars and galaxies in different wavelengths, which is something vital if one wants to understand their physics. A still better method, of course, is to use a prism – or some equivalent device – to resolve the light into various wavelengths. Such a spectroscopic study contains valuable information about the characteristics of the astronomical objects.

The traditional photographic plate has been challenged in the last few decades by a still more powerful technique, based on what is called a Charge Coupled Device (or CCD). The CCD is just an oversized silicon chip containing an array of small light-sensitive regions called pixels. The latest CCDs can have more than a million pixels arranged in a square array of 1024×1024 pixels. When the image from a large telescope is focused on the silicon slab, which is only a couple of centimeters across, it builds up an electric charge at each of the pixels proportional to the brightness of the light falling on it. After the exposure, the charges can be read off into a computer memory and can be analyzed at leisure. CCDs are extremely efficient and can respond to about 75 percent of the light falling upon them. Contrast this with the photographic plates of the 1940s, which could record only about $(1/300)$th of the light falling on them. Hale designed the 5 meter telescope to be four times more sensitive than the 2.5 meter one; but with a modern CCD he could have achieved the same sensitivity with a 0.4 meter telescope.

3.3   Heat in the sky

Right from the days of Newton, it was known that sunlight could be separated into seven different colours by passing it through a prism. In 1800 Sir William Herschel attempted to see whether all the colours of sunlight carry the same amount of 'heat'. He spread the sunlight into its rainbow colours and placed a thermometer in each of the colours in turn. He did not find any significant difference between the heat carried by the colours. However, he found that the thermometer recorded heat even when he moved it beyond the red end. It showed conclusively that some invisible radiation carrying heat is present in sunlight, and Herschel called it 'infrared'.

Today's astronomer defines the infrared band of the spectrum as radiation with wavelengths between about 0.7 micron and 1000 microns. (One micron is one millionth of a meter.) So the infrared band is fairly wide, with the largest wavelength being nearly a thousand times larger than the shortest wavelength. (Contrast this with the optical band, in which the longest wavelength is only about twice as long as the shortest wavelength.) Obviously, exploring the skies with infrared radiation will show us a lot more structure. The difficulty, of course, is that this IR radiation is not as straightforward to detect as optical light.

The simplest infrared rays which could be detected are those which are just beyond the red light. Though the human eye is not sensitive to this range, many of the photographic techniques used by astronomers in the visible band can be used to cover this range of 0.7 to 1.1 microns. The sky does not look dramatically different at these wavelengths except for some minor curiosities – e.g., Betelgeuse, the red star in Orion, becomes the brightest star in the sky, brighter than Sirius. The major advantage of infrared photography is in probing the dusty regions of the universe. The dust particles in the universe have sizes which are typically the same as the wavelength of visible light, and hence they block visible radiation very effectively. Hence, the central regions of dense dust clouds in space are completely invisible in the optical waveband. Infrared radiation, even that close to red, can penetrate the dust and reveal the mysteries which are lying deep within.

Naturally, the sky becomes more interesting at longer wavelengths. But now the problems of detection are quite serious. Normal photographic techniques are useless at these wavelengths, and astronomers have to use special materials like lead sulphide or indium antimonide to produce sensitive detectors. These substances work quite well and can substitute for the simple photographic plate of optical astronomy. However, there is one major difference: photographic plates do not emit light, but the infrared detectors do emit infrared radiation if they are kept, let us say, at room temperatures. So in order to achieve high efficiency, the detectors have to be cooled to low temperatures. At still longer wavelengths one uses semiconductor devices like germanium and gallium, but even they have to be cooled to near-zero temperatures to maintain sensitivity.

The Earth's atmosphere is both a blessing and a curse at infrared wavelengths. It is a blessing because the atmosphere scatters infrared radiation much less than visible light. The blue of the sky, which is due to scattered sunlight, prevents us from seeing the stars during daytime. But some infrared observations can take place even during daytime. Optical telescopes which are otherwise lying idle during daytime can be put to good use by infrared astronomers.

The curse of the atmosphere lies in the carbon dioxide and water molecules, which absorb infrared radiation very strongly, thereby completely obscuring the sky. Fortunately there are narrow windows of wavelength between these molecular absorption bands – like at 2.2 microns and 10–20 microns, for example – which astronomers can make use of. There is, however, a problem in making use of the 10 micron window. It should be remembered that any material object at some temperature emits thermal radiation (see chapter 2). The strongest emission from any body will be at a wavelength which is inversely related to its temperature. The Sun, with a surface temperature of 5500 K, radiates mainly in visible light. An electric stove at 1000 K emits infrared radiation of 2–3 microns; but the sky above an astronomical observatory at high altitudes will have a temperature around 270 K and will emit strongly infrared radiation at 10 microns! So an infrared telescope operating at this wavelength will see the entire sky illuminated both day and night. The spatial and temporal variations in this background will produce a noise in infrared telescopes. One has to take special precautions to subtract out the background radiation if one has to see the infrared objects.

In spite of all these difficulties, infrared astronomers have painstakingly built up images of stars, galaxies and the outer planets at these wavelengths. There are now several special purpose infrared telescopes in the world. The Hawaii mountain 'Mauna Kea' (mentioned in the last section) has not only optical telescopes but also two of the world's largest infrared telescopes.

The problems of the atmosphere can be tackled more effectively by flying the detectors above the atmosphere in aircraft or rockets. (This is especially important in the far infrared band, in which the atmosphere absorbs almost all the radiation; one cannot see much even from the highest mountains on Earth.) The first airborne infrared observatory is called 'Kuiper'. This is a facility operated by NASA which flies a team of astronomers two or three times a week at an altitude of about 13 km in a specially constructed aircraft. There is a 91 cm telescope pointed at the sky through a large hole in the side of the plane, through which infrared observations can be made. This facility will soon be replaced by a more sophisticated one, the Stratospheric Observatory For Infrared Astronomy (SOFIA).

Better still, one could put a satellite dedicated to infrared observations in orbit. This was done in 1983 when the US, Netherlands and UK collaborated to put the Infrared Astronomy Satellite (IRAS) at an altitude of 900 kms. This was probably one of the most complicated missions of its time, especially since huge amounts of liquid helium had to be kept near zero temperatures without human intervention. (In fact, the useful lifetime of the satellite is essentially decided by how long this liquid helium can maintain the low temperature.) But the mission was a notable success. IRAS surveyed the sky at the whole range of wavelengths from 8 to 120 microns with a sensitivity hundreds of times better than balloon based detectors. The IRAS Catalogue now contains a third of a million objects – stars, gas clouds and galaxies. We shall see later how models for galaxy formation are tested by looking at the distribution of galaxies in the sky. The results of the IRAS survey played a decisive role in this investigation.

This success was soon followed up by several other projects. The Cosmic Background Explorer Satellite (COBE), launched by NASA in 1989, had the ability to do infrared observations (though it rose to fame in a completely different context, which we shall discuss in later chapters). The biggest infrared observatory currently in orbit is called the 'Infrared Space Observatory' (ISO), launched in 1995 by the European Space Agency. This will operate for at least two years, and covers the wave-band from 2.5 to 240 microns. In the next decade, NASA will be launching a Space Infrared Telescope Facility (SIRTF) which will use a 'passive cooling system' – i.e., it is arranged to radiate away its own heat rather than requiring active refrigeration.

3.4   Music of the heavens

The first serious investigation of the cosmos outside the visible band took place in the radio wavelengths during the 1930s. A new window to the sky was opened and – in fact – the spectacular growth of X-ray, infrared and ultraviolet band astronomies in later years was largely influenced by the unpredicted success of radio astronomy.

Radio waves are the longest wavelength signals which we get from the skies, and they typically span anything from a few centimeters to hundreds of meters in wavelength. The radio waves from astronomical objects are in no way fundamentally different from waves from the local radio broadcasting station. One can detect them, in principle, by tuning a receiver to the particular wavelength. There are, however, several difficulties which one has to tackle before one can successfully use these signals to probe the universe.

To begin with, the long end of the radio waves is reflected back by the upper atmosphere of the earth, called the ionosphere. It is precisely this property of the ionosphere to reflect radio waves which is used in terrestrial broadcasts to bounce the signal around the Earth. By the same token, radio waves from outer space are reflected back by the ionosphere and do not reach ground level. This difficulty can be tackled by concentrating on somewhat shorter wavelength radio waves, which is what most astronomers do.

At these shorter wavelengths, it is in principle possible to keep the radio telescopes at ground level. This by itself is a great boon to the radio astronomer when compared to his (her) colleagues, who may have to send satellites up with detectors to get cosmic information. However, the radio astronomer has to contend with increasing man-made radio noise on Earth, and should take care not to operate at the standard broadcasting wavelengths. There are international conventions which have allocated specific wavelength bands to radio astronomy, in which no terrestrial broadcasting is allowed.

The components of the simplest kind of radio telescope are the same as those of a domestic radio set. An aerial or antenna collects the energy of the radio waves hitting the telescope and converts it into electrical signals. These signals are amplified by a large factor, using amplifier circuits to produce the final output, which may be stored either in a computer or displayed in some convenient form. One usually thinks of radio telescopes as saucer-shaped objects looking into the sky. But in reality they could be in very different forms. For example, Anthony Hewish constructed a radio telescope near Cambridge in the 1960s which had 2048 wire dipoles spread over a 2 hectare field. This telescope was sufficiently sensitive to allow him to look at the 'twinkling' of radio emissions coming from the sky. His student Jocelyn Bell used this instrument to make the first major discovery in radio astronomy: she found that the emissions from some radio sources are actually arriving at very regular periodic intervals (that is, in pulses). Such a source is now called a 'pulsar', which we will discuss in chapter 4.

While the radio telescope of Hewish was very successful, a dish-shaped telescope is, by and large, more versatile. In this kind of construction, the dish-like structure reflects the radio waves to a focus at which a receiver is located. By making the telescope scan over a region of the sky, one can actually obtain a radio map of that part of the sky. Such a map is extremely useful in the study of objects which are very far away.

The fact that radio wavelengths are almost a million times longer than visible radiation has an important implication in making such detailed maps. The resolving power of any telescope – which measures the finest details which the telescope can bring out – depends on the ratio of the diameter of the telescope to the wavelength at which it is operating. A large wavelength of the radio waves implies a terrible loss in resolution. For example, a hundred meter dish working at a wavelength of 10 cm will see the sky more blurred than the human eye. Such radio telescopes are still useful for surveying large regions of the sky, but not for high resolution studies. Some of the earliest radio surveys, incidentally, led to the detection of quasars and radio galaxies, which are among the farthest objects known to us today (see chapter 7).

The difficulty of low resolution was circumvented in the 1960s by an ingenious method pioneered by the Cambridge astronomer Martin Ryle. He managed to come up with a configuration of dishes, each of which is of reasonable size, but which could together act as a dish of enormous effective area. He achieved this by using the rotation of the Earth in a suitable way – thereby beginning what is now called 'synthesis radio astronomy'. In this technique, signals are received simultaneously at a number of radio antenna separated by several kilometers. As the rotating Earth carries the radio antennas around each other, their signals are combined electronically to build a map of the sky. Effectively, one obtains a radio dish several kilometers in diameter. The most powerful radio telescope now is the Very Large Array (VLA), comprising 27 parabolic antennas each 25 meters in diameter, spread over the New Mexico desert. Even higher resolutions have been achieved by combining signals from telescopes scattered around the world in a technique called 'Very Long Baseline Interferometry' (VLBI). Designs based on similar principles have helped radio astronomers, while operating at the longest wavelengths, to achieve the highest resolution in astronomy!

3.5   The cosmic tan

The radiation which gives us a suntan and builds up vitamin D in our bodies is ultraviolet radiation, spanning wavelengths from hundreds of Angstroms to about 4000 Angstroms. On Earth we are acutely aware of its beneficial as well as harmful effects. Too much exposure to UV radiation kills the living cells of our skin. In fact, the early life forms which evolved in the protection of the ocean could only propagate into dry land because the ozone layer in the atmosphere shields the UV radiation.

Once again, what is a general blessing to humanity is an irritation to the astronomer. The ozone layer prevents most of the UV radiation from reaching the ground, and hence the ultraviolet sky cannot be explored from sea level. (There is a small band between 3100 to 3900 Angstroms which is unhindered by ozone. But this band, though technically invisible to the naked eye, is just an extension of the visible band as far as photographic plates are concerned.) In the case of infrared astronomy, we found that it is possible to do some astronomy by putting the telescope on a high mountain. But, since the ozone layer in the stratosphere is three times higher than Mount Everest, this procedure is of no use in UV astronomy.

How does one detect UV radiation? An ordinary reflecting telescope with an aluminium reflecting surface will focus the UV radiation quite well. The image at the focus can be recorded with an electronic detector or even with a photographic plate. To increase the intensity of the image, it is usual to use a device called a micro channel plate. This is made of thousands of small, very thin glass tubes stuck together side by side. Each is about a millimeter long and only a fraction of a millimeter in diameter. The front window of this system supports a metal electrode, and at the rear, a phosphorescent screen is kept. When a UV photon hits the electrode, it kicks out an electron, which is accelerated along one of the tubes by the voltage applied across it. In the process, the electron knocks about the sides of the narrow channel and produces a further avalanche of electrons. This whole stream hits the phosphorescent screen, producing a bright spot of light. This image can be recorded in a photographic emulsion or electronically scanned. The latter procedure is useful when the telescope is on a satellite which has to send the information back to Earth.

The first unmanned ultraviolet telescopes were flown on high altitude balloons, like the balloons which were used to investigate the stratosphere. Later on, astronomers used brief rocket flights to put the detector into operation beyond the ozone band for short spans of time. The most attractive procedure, of course, is to use a ultraviolet satellite orbiting outside the Earth for years on end and sending data back home.

One such ultraviolet satellite, called the 'Orbiting Astronomical Observatory' carried out the first detailed ultraviolet survey of the sky. This was followed up by several other successful satellite missions. The 'International Ultraviolet Explorer' (IUE), a collaboration between NASA, the European Space Agency and the UK, was launched in 1978 and functioned till early 1996, almost like a regular ground based observatory. At present there is another satellite, called the 'Extreme Ultraviolet Explorer' (EUE) which is in orbit. In addition to these, the Hubble Space Telescope can also do observations in the ultraviolet band.

These observations revealed a sky scattered with thousands of stars and a prominent concentration in the direction of the Milky Way. Though this is roughly the picture one would have seen in the visible wave-band as well, the details are quite different, because the stars which are picked out in the ultraviolet are quite different from the stars which shine prominently in the visible. Stars with temperatures of a few thousand degrees give out most of their light in the visible band, while stars with temperature in the range of 10 000 to 100 000 degrees dominate the UV sky. Since ultraviolet telescopes pick out the hottest stars, they also pick the youngest of them. This is because heavy stars burn much more prominently, reach very high temperatures, but also consume their fuel so quickly that they die very young. The stars which are picked out in the UV band are necessarily young for of this reason.

The fact that UV astronomy selects young stars also means that we are seeing these stars close to their place of birth. Older stars – which have been around for a longer time – would have also wandered further away from their place of birth. Thus, UV astronomy allows us to probe the regions close to star births in a somewhat indirect way. (We saw earlier that infrared astronomy can actually see the birth of stars inside dusty regions.) The UV images of spiral galaxies, for example, show the spiral arms prominently since these are the sites of star formation.

In addition to surveying the sky, UV astronomy can also provide us with valuable clues regarding the dynamics of certain astrophysical processes. Two of the most common elements in the universe, carbon and nitrogen, for example, do not produce useful spectral lines in the visible but shine in the ultra-violet band. By studying the UV spectra of these elements, one can determine their abundance at astronomical sites. This procedure has been used to study supernova explosions (see chapter 4) and in particular to estimate the amount of carbon or nitrogen thrown out by the star. Similarly, in the study of inter-stellar gas, one obtains important information from UV astronomy because the UV radiation is absorbed by the interstellar gas in selected bands.

3.6   X-raying the universe

Moving on to still shorter wavelengths (and higher energies) compared to ultraviolet radiation, we reach the region of X-rays. By convention, one takes X-rays to have wavelengths between 0.1 Angstroms and 100 Angstroms. But, as usual, the boundaries are not sharply defined – the region of the spectrum with a wavelength of hundreds of Angstroms may be called soft X-rays or extreme ultraviolet; similarly the region with a wavelength of about 0.1 Angstroms merges smoothly with gamma rays.

In spite of the high energy of X-rays, they cannot penetrate through the blanket of the Earth's atmosphere to reach sea level. This, of course, is essential for the survival of life forms on Earth since most of them cannot withstand sustained exposure to X-rays. But as usual, this makes life difficult for the astronomer, who has to rely on balloons, rockets and satellites to probe the X-ray sky.

One may say that X-ray astronomy came of age on 18 June 1962 when an Aerobee rocket, launched from New Mexico, flew for a few minutes well above the atmosphere before falling back to Earth. The rocket contained detectors supposed to see X-rays emitted by the Moon's surface when high energy particles hit the Moon. Though scientists were hoping that the detector would see other X-ray sources in the sky, it was not a popular idea at that time. But this hope was indeed realized! The X-ray detectors hardly saw the Moon or the bright stars or any of the conventional objects familiar to us from optical astronomy. Instead, it saw a dim glow of X-ray background all around the sky, and a brilliant source of X-rays in the constellation of Scorpio at a position where there is no bright star!

Since then, X-ray astronomy has been contributing significantly to our knowledge of the universe, especially since the X-ray view of the sky comple-ments the other bands very successfully. The primary reason for this has to do with the extreme high energies available in the X-ray band. As we have seen repeatedly, the energy of photons increases with decreasing wavelength. To produce such energetic X-ray photons in large numbers by *thermal* radiation, the material has to be extremely hot. For example, to produce X-rays of about 10 Angstroms wavelength will require a temperature of about a few million Kelvin. Thus, only extremely hot gas can produce X-rays by thermal radiation.

For this reason, a simple star like the Sun looks very different in X-rays compared to optical wavelengths. (In fact, X-ray observations of the Sun had been carried out in brief rocket flights even in the 1950s.) The disc of the photosphere which we see every day in the Sun is not hot enough to emit X-rays. It appears as a black glow in X-ray photographs, surrounded by gas at a million degrees or so making up the solar 'corona'. If all the stellar sources were like our Sun, then it would have been extremely difficult to detect any X-radiation from the stars. However, there are other processes which can lead to emission of X-rays. For example, the Scorpio X-1 detected in the rocket flight mentioned above consists of a system of double stars – one normal star and one compact star in orbit around each other. The gas from the normal star swirls around the compact companion as the stars go around each other. This gas stream heats up due to friction and acts as a source of X-rays. X-rays can also be emitted by subatomic particles like electrons travelling in magnetic fields or interacting with lower wavelength photons.

The high energy of the X-rays also requires special kinds of detectors. The earliest detectors were somewhat refined versions of what are known as 'Geiger Counters'. These counters contain two wire grids kept inside a chamber filled with a gas like argon. A high voltage difference is kept between the two wires. When an X-ray photon enters the chamber, it knocks out the electrons from the argon atoms. These electrons are accelerated by the potential difference between the two wires and knock out more electrons from the atoms. This avalanche of electrons makes up a tiny current which is picked up, amplified and detected. The strength of the current is proportional to the number of electrons ejected, which in turn depends on the energy of the initial photon. Thus these counters can actually measure the energy of the original X-ray photon.

While this detector is satisfactory in some respects, it is not much by way of a telescope. To make an X-ray telescope one needs to focus the X-rays, which is an extremely hard task. Lenses and conventional mirrors are more likely to absorb the X-rays than bend them. To build an X-ray telescope, astronomers had to use extremely ingenious techniques. The basis for this was provided by the X-ray astronomer Riccardo Giacconi, who constructed a metallic cylinder with a special tapering surface. The shape was so adjusted that X-rays hitting the detectors at near grazing angle will be reflected to another point where the angle will still be grazing. This process continues to reflect the X-rays, eventually leading to focusing. More sophisticated techniques and the accompanying electronics have made present day X-ray astronomy capable of giving detailed maps of the X-ray sky. These detectors have been flown in several special purpose X-ray satellites like UHURU, the Einstein observatory and ROSAT. Of these, ROSAT, a joint venture between the United States, Germany and the UK launched in 1990, is still in operation. Other satellites currently in orbit include the 'Beppo SAX', an Italian–Dutch satellite launched in 1996, the 'Rossi X-ray Timing Explorer' (RXTE) which was launched at the end of 1995 and is capable of making very precise timing measurements of X-ray objects, and the 'Advanced Satellite for Cosmology and Astrophysics' (ASCA) which is a joint US–Japanese venture, launched in 1993.

It is amusing to note that X-ray astronomy developed much more rapidly than its optical or radio counterparts. In optical astronomy, it took three centuries to reach the high resolution limit starting from Galileo's telescope. Radio astronomers took four decades for the same, while in X-rays it was probably 16 years from 1962 to the launching of the Einstein Observatory in 1978.

3.7   Gamma-ray astronomy

In the last few sections we scanned through the various bands of radiation which astronomers use to probe the skies. We have now reached the highest energies, which occur in gamma-ray astronomy – a research area which has attracted considerable amount of attention in recent years.

Gamma rays have wavelengths which are shorter than one-tenth of an Angstrom. One gamma ray photon with wavelength of, say, 0.01 Angstrom will carry an energy of a few million electron volts. This is the range of energies encountered in typical nuclear phenomena. In fact, the mass of the electron – the lightest known elementary particle – is equivalent to an energy which is about half a million electron volts. Thus, gamma ray has higher energy than an electron, and it is indeed possible to produce electron–positron pairs out of gamma rays under suitable conditions. To emit a significant amount of gamma rays by thermal radiation, the body has to be heated to a temperature of more than ten billion degrees! Needless to say, there are no objects in the universe which are steadily maintaining such high temperatures. Thus, detection of gamma rays from the depths of space will indicate an essentially nonthermal mechanism for their production. That opens up an exciting array of theoretical possibilities which interest astronomers.

Just like many other kinds of radiation, gamma rays are also strongly absorbed by the Earth's atmosphere (which is nice, since they are harmful to living organisms). So, to detect gamma rays, one has to use satellites or similar probes. The high energies of the gamma rays allow them to pass right through the gas in simple proportional counters or other detectors used at longer wavelengths. So gamma ray astronomy also requires special techniques for detection. The most versatile detector for these wavelengths is called a 'scintillation counter' and operates in the same way as the luminous dials in wrist watches. The essence of a scintillation counter is a large crystal of sodium iodide, surrounded by photomultiplier tubes. A gamma ray hitting the sodium iodide loses a major fraction of its energy, which is emitted in the form of ordinary light. This light can be picked up by the photomultiplier tubes and amplified. Brighter flashes of light will be produced by more energetic gamma rays; by measuring the energy deposited in the photomultiplier tubes, one can estimate the original energy of the gamma ray. If enough gamma rays are gathered from a single source, one can even attempt to obtain the spectrum of the source.

This was roughly the way the earliest satellites – in the 1960s – attempted to probe the gamma ray band. The preliminary results showed that there is gamma radiation coming from the Milky Way region in the sky. However, these satellites could not give directional information about the gamma rays – so one did not know where they came from and could not 'photograph' the gamma ray sky. To do that, one requires a gamma ray telescope. Once again, we face the problem of focusing the radiation. In the case of X-rays, one could just about manage to do it by grazing-angle techniques; for gamma rays, even this method fails. One just cannot use any known substance as a mirror which could focus gamma rays.

It is, however, possible to get the directional information of gamma rays by another method. A kind of gamma ray 'telescope' can be constructed by stacking together a series of detectors called spark chambers interleaved with thin tungsten sheets. Spark chambers are capable of recording the tracks of electrons and positrons which pass through them, and are widely used in nuclear and particle physics research. A gamma ray (with wavelength shorter than 0.001 nanometer) that enters such a telescope will pass through the stacks of

63

spark chambers and tungsten sheets until it passes close to a tungsten nucleus. When that happens the gamma ray photon is likely to convert its energy into an electron–positron pair. The pair continues to travel through the stack and reaches the bottom of the telescope, where a scintillation counter measures its energy. The track of the electron–positron pair is recorded in the spark chambers. Knowing these, one can reconstruct the original direction in which the gamma ray was travelling – or, equivalently, the direction of the source in the sky.

Such gamma ray telescopes were used in the American satellite SAS-2, which was flown in 1972, and the European satellite COS-B. The COS-B telescope had extremely poor resolution by the standards of any other band – it was about two degrees, which is four times the size of the Moon in the sky. Even though a source could be located only within such a broad area, COS-B produced the first gamma ray map of the sky. The Sun does not emit many gamma rays – except during a flare – and the gamma ray maps are dominated by the Milky Way. It is believed that these galactic gamma rays originate when high speed charged particles present in cosmic rays hit the gas atoms in the Milky Way. In that case one expects more gamma rays to come from the regions where the gas density is higher – and this is indeed what COS-B found. The satellite also found two prominent flashing sources in the sky: the Crab and Vela pulsars. (We shall say more about pulsars in chapter 4, section 4.4.) More recently, information about the gamma ray sky came from the 'Compton Gamma Ray Observatory' (CGRO) which was launched by the space shuttle in 1990.

Gamma ray astronomy has revealed enigmatic phenomena called 'gamma ray bursts', which are perhaps the highest energy burst phenomena known. These are events in which a source emits a tremendous amount of gamma ray energy in a short period of time. The bursts last from a few seconds to several minutes, and about 200 events occur in a year. During the burst, the energy output is larger than any other gamma ray source in the sky. Because of the low resolution of the gamma ray telescopes, it is difficult to identify the source with any other known object in the sky. Observations suggest that gamma ray bursts are distributed remarkably uniformly over the sky and are likely to be from sources outside our galaxy. Astronomers are still unsure as to what causes these bursts. This is an active area of research today.

4

Getting to know the universe

4.1 Stars: Nature's nuclear plants

It is said that a man in the street once asked the scientist Descartes the question: 'Tell me, wise man, how many stars are there in heaven?' Descartes apparently replied, 'Idiot! no one can comprehend the incomprehensible'. Well, Descartes was wrong. We today have a fairly reasonable idea about not only the total number of stars but also many of their properties.

To begin with, it is not really all that difficult to count the number of stars visible to the naked eye. It only takes patience, persistence (and a certain kind of madness!) to do this, and many ancient astronomers have done this counting. There are only about 6000 stars which are visible to the naked eye – a number which is quite small by astronomical standards. The Greek astronomer Hipparchus not only counted but also classified the visible stars based on their brightness. The brightest set (about 20 or so) was called the stars of 'first magnitude', the next brightest ones were called 'second magnitude', etc. The stars which were barely visible to the naked eye, in this scheme, were the 6th magnitude stars. Typically, stars of second magnitude are about $2\frac{1}{2}$ times fainter than those of first magnitude, stars of third magnitude are $2\frac{1}{2}$ times fainter than those of second magnitude, and so on. This way, the sixth magnitude stars are about 100 times fainter than the brightest stars. With powerful telescopes, we can now see stars which are about 2000 million times fainter than the first magnitude stars, and – of course – count them.

In comparing the brightness of the stars, as seen from Earth, one should keep the following fact in mind. A star might appear faint either because it is intrinsically less bright or because it is at a large distance. So we can really compare the intrinsic brightness of two stars only if we know their relative distances from us. Alternatively, we can compare the distances if we know the intrinsic brightness of the stars.

But how can we find the intrinsic physical properties of a star which is so far away? The answer lies in the analysis of the spectrum of the starlight.

The light we receive from the star contains a mixture of electromagnetic radiation of different wavelengths. For example, the visible light we receive from the Sun can be split into different colours (corresponding to different wavelengths) by passing it through a prism. By similar – but more sophisticated – methods we can split up the light from the star into various wavelength bands. Stars differ widely in their spectrum and, by studying the spectrum, one can understand the intrinsic properties of the star to a great extent.

We saw in chapter 2 that objects kept at a particular temperature will emit thermal radiation with a characteristic spectrum, shown in figure 2.14. It is clear from this figure that most of the radiation is emitted at a frequency which increases with increasing temperature. The radiation which we receive from stars looks like a thermal spectrum to a fair degree of accuracy. Most of the radiation emitted by a star will be at a frequency determined essentially by the surface temperature of the star. If the maximum radiation is around the wavelength corresponding to blue colour, the star will appear bluish in the sky like Spica (in Virgo) or Bellatrix (in Orion); if it corresponds to the wavelength of red colour, the star will appear reddish like Betelgeuse (in Orion) or Antares (in Scorpio). Since blue has a higher frequency than red, bluish stars are significantly hotter than the red ones. Study of the spectrum shows that the surface temperature of stars varies considerably from star to star. The blue stars can be as hot as 30 thousand degrees, while the surface temperature of a star like the Sun is about 6000 degrees. Because of the connection between colour and temperature, astronomers often use the colour of the star as an indicator of its temperature.

In addition to the thermal radiation, the spectra of stars will also show radiation emitted by individual atoms which are produced when the electrons in the atoms jump between different levels. This radiation will occur at definite wavelengths and is called 'emission lines'. These lines contain valuable information about the chemical composition and physical conditions that exist in the star. In particular, by diagnosing the various spectral lines present in the light emitted by a star, one can obtain information about the chemical elements in the star (see figure 4.1). It turns out that stars are mostly made of hydrogen and helium with small traces of other heavier elements. As we shall see in the next chapter, this result is of major cosmic significance. In the least, it shows that stars have a very different chemical composition compared to, say, that of Earth.

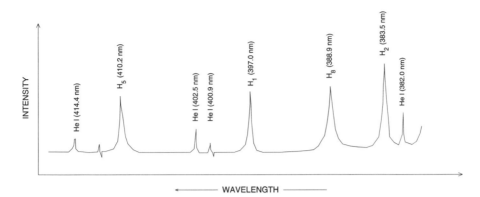

Figure 4.1 Analysis of the spectrum of a star provides valuable information about its properties. The spectrum (like the one shown here) shows characteristic peaks at those wavelengths which correspond to transitions between the electronic levels in the atoms. Since different atoms have different electronic levels, these peaks will allow us to identify the composition of stellar material. Studies show that stars are mostly made of hydrogen and helium. The spectrum shown here has prominent emission lines of hydrogen and helium at different wavelengths.

Another important parameter which characterizes a star is its 'luminosity'. Luminosity measures the total amount of energy emitted by the star per second. Loosely speaking, we would expect more massive and bigger stars to emit more energy. Hence the luminosity of a star can depend on both the mass and the radius of the star. We thus have two observable parameters for any star, viz. its surface temperature and luminosity. Both these can depend on the mass, radius and several other properties of the star. In general, therefore, one will not expect to find any simple relation between the luminosity and surface temperature of a star. Incredibly enough, such a relation exists! To look for such a relation in a systematic manner, one could plot the location of all the stars in a diagram, giving their luminosity and surface temperature. Such a diagram, called a 'Hertzsprung–Russell diagram', reveals a remarkable feature. A schematic version of this diagram is given in the left frame of figure 4.2. (We saw earlier that the colour of the star is related to its temperature. The temperature is shown along the top horizontal axis and the colour along the bottom horizontal axis.) A priori, one would have expected to see stars scattered all over the diagram. That is, one would have expected to see stars with all kinds of temperatures and luminosities. But figure 4.2 shows that most of the stars lie on a diagonal band from the top left to the bottom right. This band is called the

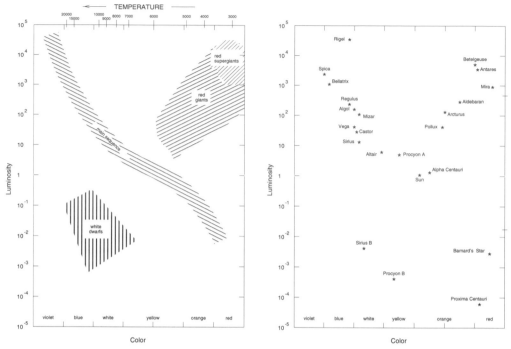

Figure 4.2   By studying the spectrum of light emitted by a star, one can estimate its temperature and total luminosity. In general, one would not have expected to see any relation between these two quantities. So, if we mark the position of different stars in a diagram by giving their luminosity and temperature, we would expect the stars to be randomly distributed all over the diagram. In reality, however, most of the stars fall in a thin diagonal band from top left to bottom right in such a figure. The stars in this band are called 'main sequence' stars and they have a definite relationship between the luminosity and the temperature. This is shown schematically in the figure on the left. (The regions marked 'white dwarfs', 'red giants' and 'red supergiants' are discussed in the text later.) Some familiar stars are shown in the figure on the right side. The surface temperature of the star is indicated on top of the picture (note that the temperature increases to the *left*) and the colour of the star is denoted at the bottom.

'main sequence'. Clearly, the luminosity and temperature of the main sequence stars are related to each other in a definite manner. The right frame of figure 4.2 gives the location of several familiar stars in the diagram. Of course, there are far more stars along the main sequence than indicated in this frame.

Why are most of the stars clustered in the main sequence? This is essentially because an average star spends most of its lifetime in the main sequence. So when we select a whole bunch of stars from the sky, we have a greater chance of catching the stars in the main sequence than outside it. Hence, the first thing we need to understand about a star is what keeps the star in the main sequence, shining for billions of years together. The Sun, for example, emits about 4×10^{33} ergs of energy in one second. Equivalently, every square meter region of the Sun's surface is emitting a power of 62 000 000 watts. Where does this supply of energy come from?

This had been a source of mystery for scientists till the early part of this century. It was fairly easy for astronomers to realize that the conventional sources of energy will not do. For example, if the Sun is made of best quality coal which is burning – and assuming oxygen is being supplied! – the energy output can only last for about 1500 years. This shows that conventional chemical fuels are quite inadequate. More realistically, one can imagine the possibility of gravitational energy being converted into heat. If the Sun contracts under its own weight, the gas molecules making up the Sun will be compressed to higher pressure and temperature and will provide a source of heat. Assuming the Sun was a very large sphere of gas to begin with, its contraction to its present size would have maintained the energy supply for about 20 million years. That might seem a long time by everyday standards but it is not long enough. Geologists have found evidence that the Earth's crust has existed, and was illuminated by the Sun, for a much longer period.

The reason the Sun – and the other stars – have been shining for such a long time is that they use one of the most efficient sources of energy – viz. nuclear power. In nuclear reactions the mass is directly converted into energy. Such a process is nearly a million times more efficient than the best chemical fuel. This is the secret of star power and leads to a fairly complicated life cycle for the stars. To see it in action, let us follow the evolution of the star from its birth to its death.

4.2   Story of a star

According to current astrophysical models, we may describe the story of a star along the following lines. Consider a large, massive, gas cloud which begins to contract under its own weight. The gravitational contraction of such a cloud will compress the material and cause its temperature to increase. Eventually the central temperature of such a cloud can be as high as 10 million Kelvin. At such temperatures, hydrogen would be ionized and the gas will exist in the plasma state as a bunch of ions and electrons. All these particles will be moving at very high speeds and colliding with each other frequently. At such high energies, the collision of hydrogen nuclei can fuse them together through the process of nuclear fusion. We saw in chapter 2 that nuclear fusion of light elements can release enormous amounts of energy. Such a reaction occurs only in the central regions of the gas cloud; in the Sun, for example, only about 12 percent of the mass is involved in nuclear fusion. But the release of nuclear energy by such a process is enormous. It is this energy which flows out in the form of radiation which we eventually see as starlight.

The actual nuclear processes which take place in a star are somewhat complicated and involve several chains of reactions; but the principle is the same. Lighter nuclei are fused together to form heavier elements, releasing energy in the process. In fact, it is the triggering of this nuclear power which signals the birth of a star. The first nuclear reaction which occurs copiously involves fusing hydrogen to form deuterium and then helium-3 or helium-4. This process can provide nuclear energy to keep the stars shining for a long, long, time. In the case of our Sun, the original contraction of the gas cloud lasted about 15 million years, after which the nuclear reactions were ignited. It has already lived on the energy released by the nuclear reactions for nearly 4.6 billion years. Throughout this time there has been very little change in the structure of the Sun. In fact, nearly 90 per cent of all the stars we see in the sky are in such a phase; the Pole star, Sirius (in Canis Major), Vega (in Lyra) and the three stars in the belt of Orion are some of the more familiar examples of what the astronomer calls the 'main sequence stars'.

This phase ends when the hydrogen in the core of the star is exhausted – which will happen to our Sun in another 5 billion years. A star like the Sun would have by then spent nearly 10 billion years in the main sequence. This time span, of course, is not the same for all the stars. The more massive a star, the more will be the gravitational contraction during its formation and the higher will be its central temperature. This means that more massive stars will 'burn' the hydrogen faster and will also have higher luminosity. For Sirius this phase will scarcely last more than 100 million years, and for some of the stars in Orion it will be less than 10 million years. Sooner or later the star has to come to terms with the fact that hydrogen burning cannot keep it shining for ever.

At the end of hydrogen burning, the stellar core is mostly made of helium. As the star contracts further, under the action of gravity, the temperature soon crosses 100 million degrees at the centre. Around this stage the next key transformation occurs. It turns out that the total mass of three helium nuclei is almost equal to the mass of an excited state of a carbon nucleus. (Just as an electron in a hydrogen atom can exist in ground state or in excited energy state, the nuclei can also exist in different energy states; since mass and energy are interconvertible, we can think of an excited nucleus as having a higher mass.) Hence, by fusing three helium nuclei together, one can produce carbon, release more nuclear energy and proceed to the next stage in the synthesis of elements. (It is remarkable that the *necessary* existence of such an excited state of carbon was theoretically predicted by Fred Hoyle before it was seen in the laboratory! It was one of a dozen or so occasions in physics when a physicist showed deep insight into the workings of nature without the aid of experimental data.) The starting of this nuclear reaction affects the entire structure of the star in a complicated manner. The increase in the pressure and the temperature in the core could eventually push the outer envelope of the star to expand to a very large size. This expansion lowers the temperature of the star somewhat and it ends up as a 'red giant', giant because of its large size and red because of its colour. Betelgeuse (in Orion), Aldebaran (in Taurus) and Antares (in Scorpio) are examples of red giants.

Further nuclear reactions can ensue, starting from carbon. For example, carbon can combine with helium to give birth to oxygen, and later on, through a relatively complex series of reactions, yield neon, sodium, magnesium, aluminium, silicon, etc. This way the star can build up heavier and heavier elements by nuclear fusion, as long as nuclear fusion leads to a release of energy. A star which is about fifteen times as massive as Sun can slowly evolve into a structure containing concentric shells of different elements like hydrogen, helium, carbon, neon, oxygen, silicon and iron. Stars, therefore, act not only as nuclear reactors but also as laboratories in which various heavier elements are synthesized from hydrogen.

To model any such star mathematically, one needs to invoke mathematical relations from different branches of physics. The equilibrium of a gaseous sphere is maintained because the effect of gravity trying to contract the gas is counter-balanced by the gas pressure. The energy transport from the central regions (where nuclear reactions take place) to the surface of the star is governed by the properties of the gas outside the nuclear core. Given the equations governing these processes, it is possible to devise a complete model for the main sequence star. Such models can be used to relate the luminosity of the star to its surface temperature. It is the existence of such a relation which makes the stars cluster along a diagonal band in figure 4.2.

You might have noticed that we started with a gaseous cloud of hydrogen, formed stars and synthesized heavier elements inside the stellar crucible. Why did we do this? Couldn't we have started instead with a gaseous cloud of, say, oxygen or neon and proceeded further? This question is fundamental and the answer has to do with cosmology. According to current cosmological models, the universe is continuously evolving and was considerably hotter in the past. We shall see in the next chapter that when such a universe cools, it essentially will be made of hydrogen and helium with very little of the higher elements. The cosmological models predict that the initial gas cloud will be made of about 76 percent of hydrogen and 24 percent of helium. Hence, the initial, first generation, stars *have* to form from a cloud of hydrogen and helium. In fact, we owe the existence of all higher elements to stellar nucleosynthesis.

4.3   Supernovae: the violent end

The scenarios described in the last section give a qualitative picture of the rich variety of phenomena which are possible in stellar evolution. In general, much more complicated scenarios involving several combinations of nuclear processes are possible. But whatever the intermediate process, it is inevitable that the nuclear energy supply will come to an end at some stage. When this happens, gravity will make the star contract further and make it denser and denser. If the total mass of the star is not too high, this contraction can be halted when the electrons in the star are squeezed to very high densities. The squeezing of the electrons provides a pressure (called 'degeneracy pressure') which could stop further contraction of the material. Such an end product of stellar evolution is called a 'white dwarf'. A white dwarf can have a radius which is comparable to that of the Earth but a mass which is about that of the Sun.

The final fate of a star which has significantly higher mass can be more spectacular. During its last stages of evolution, it is capable of ejecting its outer envelope in a powerful explosion. When this happens, the star suddenly becomes very luminous and is called a 'supernova'. The 'nova' in supernova stands for a 'new star' and suggests the discovery of a bright new star in a location in the sky. In a way, this is a misnomer; it is not true that there was no star there before. What usually happens is that an old star suddenly brightens up in a span of a few days, and – later on – slowly fades away in a month or two. The brightness of a supernova increases by a factor of several hundred million in a matter of days, and it can stay as such a bright object for a few weeks or months. During the brightest phase such a star can outshine all other stars in that constellation and even change the shape of the constellation completely.

It is estimated that there is one supernova explosion occuring somewhere in the universe every second; but, of course, most of them are too far away to be visible from the Earth. Supernovae in our galaxy or in nearby ones (which are bright enough to be seen by the naked eye) are estimated to occur only about once in a century; in that sense, it is a rare event. Two of the historically important supernovae occurred in 1572 and 1604. The first one occured in the constellation of Cassiopeia and was studied by the Danish astronomer Tycho Brahe. The supernova was bright enough to distort significantly the familiar 'M' shape of that constellation. The second one happened to occur near a conjunction of Mars and Jupiter and thus was noted by astrologers all over the world. The systematic study of this supernova was made by the astronomer Kepler. In the last millennium, only half a dozen supernovae occured which were bright enough to be seen with the unaided eye, and only one of them occurred after the invention of the telescope – in 1987! This occurred at the location of a massive star called Sk-69 202 in a nearby galaxy called the Large Magellanic Cloud (see section 4.8). As the supernova faded, it was clear that the star Sk-69 202 has vanished. Thus, for the first time, one saw a supernova arise from the explosion of a massive star.

A supernova can leave a tell-tale mark around the local region in which the explosion took place. Though the region between the stars – called the interstellar medium – is more tenuous than the best man-made vacuum, it is not completely empty; it has about one hydrogen atom per cubic centimeter. The explosive outflow of material from the supernova will sweep up the interstellar matter until the accumulating pressure can eventually bring the outflow to a halt. A supernova, thus, creates a hole in the interstellar medium surrounded by a shell of compressed matter. These objects – called supernova remnants – last for tens of thousands of years and can be seen in optical, X-ray and radio bands. A recent radio survey has seen more than a hundred such remnants in our galaxy; when these regions are studied through an optical telescope one often finds a wisp-like nebula in the vicinity. The first supernova remnant to be discovered was the Crab nebula in Taurus, which was found at the precise location of a supernova observed by Chinese astronomers in 1054. Several other supernovae, like Kepler's and Tycho Brahe's, have likewise been identified with remnants.

Supernova explosions are not only spectacular but also provide an extremely important service. We saw earlier that heavier elements can be synthesized inside the stars, starting from hydrogen and helium. But we do see different elements in the Earth. If this element synthesis took place inside the stars, how did they get out of the stars? How did Earth, for example, end up having these heavier elements? The answer is provided by supernova explosions, which spew out the heavier elements synthesized in the star into the interstellar medium. (The supernova also synthesises some of the heavy elements, but we shall not discuss this phenomenon.) Planets, as we understand them today, have condensed out of this medium which contains heavier elements.

This process also injects a new dynamism into the birth and evolution of stars. Notice that in the early life of the galaxy stars must have formed out of a mixture of hydrogen and helium. Thanks to the first round of supernova explosions, the galaxy acquired a medium enriched by heavier elements including carbon. Later generations of stars are born out of this enriched gas and will contain a small population of heavy atoms right from the beginning. It turns out that the presence of even a small amount of heavy atoms can influence stellar evolution quite significantly. Carbon, for example, has the effect of increasing the rate of flow of energy in the star, thereby shortening the life of the star; this, in turn, leads to quicker dissemination of heavy atoms into the interstellar medium. Thus, in a way, carbon can increase its own rate of formation. Such processes show that the second (and later) generation of stars can have a more complicated life history compared to the first generation stars.

4.4   Neutron stars and pulsars

What happens to the core of the star when the supernova explosion throws away the outer envelope? Once again the answer depends on the mass which remains in the core. If this mass is less than about two solar masses, the core ends up as a 'neutron star'. The rapid collapse of the core of the supernova squeezes the protons and the electrons together, forming neutrons, thereby producing a remnant which is made of neutrons. The density of such a neutron star will be comparable to that of an atomic nucleus, which is abut 10^{15} gm cm^{-3}. This means that a match-box full of neutron star material will weigh 1000 billion kg!! Neutron stars are tiny objects having a radius of *only* 10 km, which is 600 times smaller than the radius of the Earth.

The truly remarkable features of the neutron star are its magnetic field and rotation. All the stars (including our Sun) rotate, but only very slowly. As the star collapses to form a compact remnant, its speed of rotation increases. (This is for the same reason that an ice skater revolves more quickly by bringing his arms closer to the body.) A compact neutron star can rotate several hundred times in a second, compared to our Sun, which takes about 27 days to rotate once on its axis. All stars possess magnetic fields as well. As the star collapses, the strength of the magnetic field also increases, since it gets confined to a smaller region. The neutron star will have a very high magnetic field on its surface which could be a million million times that of Earth.

These two properties – high rotation and magnetic field – allow the neutron stars to manifest themselves as a special kind of stellar object called 'pulsars'. The first pulsar was discovered in 1968 by Hewish and Bell at the Mullard observatory of Cambridge. They found a celestial object which was emitting radio pulses of varying intensity but with an extremely regular spacing. The period at which the signals were received was incredibly precise – 1.33730113 seconds. (The precision was so high that it was even suggested that these could be radio *signals* coming from extra-terrestrial beings!) The emission arises from a rapidly rotating neutron star. The magnetic field of the neutron star accelerates the electrons around the neutron star, causing them to emit beams of radio waves. The pattern of emission rotates along with the rotation of the star and reaches us only when the direction of emission is towards the Earth. Thus we get the pulsar radiation at very regularly spaced intervals.

Since 1968, several more pulsars have been discovered, and we know today of more than 700 pulsars. Some of them emit radiation not only in the radio band but also at higher frequencies like X-rays and gamma rays. As the pulsar radiates energy, its rotation slows down at a very small rate. Such a decrease, called the 'spin down rate', can be measured by careful monitoring of the pulsar over years. By the same token, one would have expected young and recently formed pulsars to rotate much more rapidly compared to older ones. This is indeed true; the pulsar in the Crab nebula, formed out of the supernova explosion in 1054 AD, is one of the most rapidly rotating pulsars known. (There are some exceptions to these general rules; we shall not discuss such complications.)

Another remarkable object belonging to the same class is a 'binary pulsar', first discovered in 1974 and called PSR 1913 + 16. (These numbers denote the position of the object in the sky.) This system consists of a 1.4 solar mass neutron star pulsing nearly 17 times a second, accompanied by another invisible neutron star of similar mass. These two compact stars rotate around each other in a very tight orbit with a period of about 8 hours. This binary pulsar has been used as an interesting laboratory to test the predictions of the general theory of relativity. We saw in chapter 2 that, according to Einstein's theory, the gravitational field should be properly viewed as curvature in space. The predictions of Einstein's theory differ from the simpler version of gravity described by Newton. But the differences are so tiny that it is fairly difficult to test them in terrestrial laboratories. For example, one of the predictions of Einstein's theory is that rapidly accelerating material bodies should emit energy in the form of gravitational waves. This phenomenon does not exist in the Newtonian theory of gravity, and the detection of gravitational waves will be an important landmark in testing Einstein's theory of gravity. The loss of energy due to this process will lead to a gradual decrease in the orbital size of a pair of bodies revolving around each other. This effect, in principle, should exist even for planets revolving around the Sun. However, in the solar system the gravitational field is not strong enough to lead to a detectable effect. Neutron stars, on the other hand, have a much stronger gravitational field, and hence the change in the orbital parameters due to the emission of gravitational waves can be detectable in these systems. It turns out that the binary pulsar PSR 1913 + 16 can be used to verify the existence of gravitational waves in an indirect manner. In this binary pulsar the gravitational field is sufficiently strong for the effects of Einstein's theory to manifest themselves. These pulsars must be emitting gravitational radiation as they orbit rapidly around each other. By carefully monitoring the orbital period of the binary pulsar, one should be able to see whether it is changing as predicted by the theory. The predictions of Einstein's theory are in very good agreement with observations in PSR 1913 + 16. Its orbital period is decreasing by 76 milliseconds each year – which is tiny but detectable. In a sense the existence of gravitational waves is confirmed in this binary pulsar system, albeit indirectly.

4.5   Black holes

Does the neutron star signify the end of stellar evolution? No. The pressure due to neutrons can halt the power of gravity only if the neutron star is not too massive. There is a critical mass called the 'Chandrasekhar mass' which decides the fate of neutron stars. If the mass of the neutron star is larger than the Chandrasekhar mass, then the contraction due to gravity cannot be stopped at all. Such an object ends up as a 'black hole' – the most exotic end a star can aspire to.

Probably no other topic in astrophysics has caught the popular imagination as much as the concept of a black hole. Hundreds of popular articles, books and science fiction have been written based on the idea of a black hole. What exactly is a black hole and why does it have the esoteric properties which make it so interesting?

As a neutron star contracts, its radius decreases and the gravitational force at its surface keeps increasing. Soon a stage is reached such that no material particle can escape from its strong gravitational pull. In fact, the situation is worse than that. Not even light can escape from its surface, thereby rendering this object completely invisible for an outside observer. Such an object is called a 'black hole' and represents the final stage of gravitational contraction of a sufficiently massive neutron star.

How is it that no material particle can escape from a black hole? The correct answer requires the description of a gravitational field in terms of Einstein's theory. As we saw in chapter 2, this theory describes the gravitational force as arising due to space being curved. The curvature of space also affects the propagation of light rays. (Light rays always travel in straight lines, but the idea of a straight line on a curved surface is quite different from what we might intuitively feel.) The curvature in space produced by a neutron star can be enormous when it contracts to a very dense configuration. The path of the light rays gets highly distorted in such a strong gravitational field represented by large curvature. When the body becomes a black hole, the curvature reaches a critical value, thereby trapping the light from escaping from its surface. The critical radius to which a body must contract in order to become a black hole is called the 'Schwarzschild radius'. For a body as massive as the Sun, it is about 3 kms. This means that if the entire mass of the Sun (which is now occupying a radius of about 7 00 000 km) is compressed to a region less than 3 km in radius, then the Sun will become a black hole.

We can also look at this phenomenon in an alternative, though less accurate, manner as follows: we know that a stone thrown from the surface of the Earth will normally fall back because of the Earth's gravitational pull. It is, however, possible to shoot a rocket with such a high velocity that it escapes from the gravitational pull of the Earth. The critical speed which is needed for this, called the 'escape velocity', is about 12 km s^{-1}. For a dense neutron star this escape velocity can be much higher. When the escape velocity exceeds the speed of light we may conclude that even light cannot escape from such an object. This happens when the radius of the object is smaller than the Schwarzschild radius. (The inaccuracy in the above argument which, incidentally, has appeared in several popular articles, is the following. This argument merely shows that light emitted from the surface of such a body will eventually 'fall back' on it *after* travelling out for some distance. This is like the stone thrown from the surface of the earth which *goes up to some height* before falling back. But what happens in a black hole is quite different. The light never gets out of the surface in the first place.)

If the black holes are completely invisible, how will we ever know that they exist? Nature has conveniently provided a way in the form of star systems known as 'binaries'. These are pairs of stars bound together by gravity and orbiting around each other. If one of the stars of a binary system becomes a black hole due to gravitational collapse, the strong gravitational pull of the black hole will pull out matter from the companion star which will fall on to the black hole as a stream. Observing from Earth, we will see one star orbiting an invisible companion with a stream of matter flowing from it. By studying the orbit of the visible star, one can estimate the mass of the invisible companion. If the mass turns out to be more than, say, twice the mass of the Sun, then it is extremely likely that the invisible object is a black hole. Astronomers have identified quite a few candidates for possible confirmation as black holes in recent years.

The property of being 'black' is only one of the many curious properties of a black hole. Another feature of black holes, which is of tremendous importance theoretically, has to do with the loss of information when a black hole is formed. A normal star is described by several features like its mass, shape, density distribution, etc. Suppose this star collapses to form a black hole. The final configuration of a black hole can be described in terms of the total mass of the star alone. (This assumes that the star has no rotation and carries no electric charge. Otherwise, one needs two more parameters to describe the black hole.) In other words, two stars with entirely different physical parameters but with the same mass will be indistinguishable from each other when they become black holes. Thus a black hole gobbles up a tremendous amount of information during its formation. Related to this is the fact that the dynamics of a black hole can be interpreted using a set of laws similar to the laws of thermodynamics. For example, it can be shown that when black holes interact dynamically, their surface areas cannot decrease. This is similar to the thermodynamics dictum that during natural processes the disorder in physical systems can only increase.

In the last few sections, we have seen that the life history of a star varies considerably, based on the initial mass. A star like the Sun goes through a slow

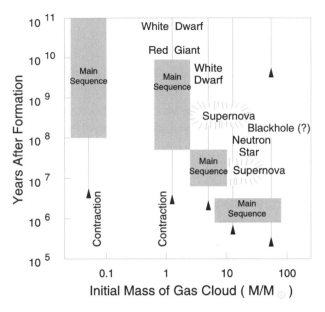

Figure 4.3   The evolutionary history of a star depends strongly on the mass of the *initial* cloud from which it is formed. A star like the Sun will go through a red giant phase and end up as a white dwarf. It is currently believed that stars below about six solar masses end up as white dwarfs, while more massive stars undergo supernova explosions, leaving behind neutron star remnants. Stars forming from more massive clouds can form black holes. At the other end of the scale, protostellar clouds with less than 8 per cent of solar mass do not ignite nuclear reactions and end up as planet-like objects.

contracting phase followed by the main sequence phase and red giant phase, and ends up as a white dwarf. Stars which are considerably more massive than the Sun are likely to end up in a supernova explosion, leaving behind a white dwarf or a neutron star. Very massive stars, on the other hand, end up as black holes. If the initial mass of the gas cloud is not high enough to ignite nuclear reactions, its end stage could be a structure like a planet. Figure 4.3 summarizes the evolutionary track of objects which start out with different initial mass.

4.6   Galaxies

The second most important unit in the cosmic hierarchy is a galaxy. In fact, cosmologists will often consider galaxies to be the basic building blocks of the universe. They show a wide variety of phenomena, and there is still a lot to be understood about the formation and evolution of these objects. Compared to stars, galaxies still remain a bit of a mystery to the astrophysicist.

To begin with, galaxies come with a wide variety of physical properties. Their masses, for example, can vary from about a million solar masses to ten million million solar masses. Those at the low mass end are called 'dwarfs' and they contain typically about a million stars. A more 'typical' galaxy, like our own Milky Way, contains about a hundred billion stars like the Sun. (As we said before, all the familiar stars we see in the sky belong to our own galaxy, the Milky Way.) At the other extreme, some of the giant galaxies contain more than 100 times the number of stars in our galaxy.

Galaxies also exhibit different kinds of morphologies or shapes. At the extremes are the two classes called spirals and ellipticals. Spiral galaxies have a disc-like appearance with distinct, spiral-shaped arms on which most of the stars reside. In contrast, elliptical galaxies are three dimensional and have the characteristic shape of an ellipsoid. The ellipticals themselves are classified into several categories depending on how distorted they are. Almost spherical systems will be called E0-type and fairly flattened ones are called E7-type. In between we have a gradation through E1, E2,..., E6. The spiral galaxies are classified according to how tightly their spiral arms are wound and whether they have a bar-like structure near their centre. Normal spirals range from S0 (which have virtually no spiral arm) through Sa, Sb and so on, which have well-developed spiral arms. Barred spirals are spiral galaxies which have a bar-like structure near the centre, with spiral arms emanating from the ends of the bars. They are again classified as SB0, SBa, SBb, etc. (see figure 4.4). Nobody has really understood the significance of these shapes from fundamental physics, and hence they merely serve as a classification scheme at present. Those galaxies which do not fit in this scheme of things are called, somewhat derisively, 'irregulars'.

(a) Spiral galaxies

Since spiral galaxies are among the most spectacular objects in the sky, it is probably best to start our discussion with them. The most important question which springs to mind regarding spiral galaxies is, of course, how the spirals could have formed in the first place. You might think that if the stars rotate differently at different distances from the centre, then they will get 'wound up'

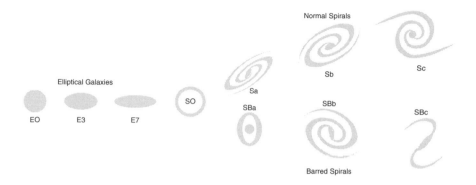

Figure 4.4 This schematic diagram shows the variety of shapes encountered among the galaxies. The elliptical galaxies are classified as E0, E1, ..., E7 depending on their ellipticity. The spirals are first divided into one of two categories depending on whether they host a bar like structure in the centre (barred spirals) or not (normal spirals). Within each of these categories they are classified as S0, Sa, Sb, ... depending on how pronounced the spiral shape is.

over a period of time, forming spiral structures. Calculations, however, show that this is *not* the cause for the spiral structures which we see. The real mechanism is actually more subtle and is described by what the astronomers call a 'density wave theory'. The idea behind the density wave theory can be illustrated by a simple example of two-lane traffic. Think of a two-lane track along which a large bunch of cars is moving at about the same speed, say 100 km hr^{-1}. Let us suppose there is an inconsiderate truck driver crawling along at 60 km hr^{-1} in one of the lanes in the middle of the fast-moving cars. This will cause a bottleneck behind the truck, since the faster cars are forced to drive in the single fast lane. This congestion behind the truck moves at the speed of the truck. If we could view this region from a helicopter at night, seeing only the headlights of the cars, the bottleneck region will appear far more luminous because of the higher density of cars. This bright spot will travel at the speed of the truck. But note that each car manages to get past the bottleneck sooner or later and resume its motion at 100 km hr^{-1}. So the bright region is due to the existence of some cars behind the truck though it is never the *same* set of cars. Clearly the bottleneck is a stable perturbation in the smooth stream of cars caused by the existence of the truck.

It turns out that spirals are caused by a cosmic traffic jam in which the role of the truck is played by a perturbation in the local gravitational force and the cars are just the stars moving in the galaxy. When the stars cross a region which has a perturbed gravitational force, their motion is slowed down and the local density of stars is increased, thereby causing a bright spot. Detailed calculations show that this could lead to the spiral shape seen in the galaxies.

Such an enhancement in the density rotating through the galaxy at a slow speed is called a 'density wave'. Density waves can also affect the gas in the disc of these galaxies and cause them to be compressed further. This compression and an enhancement of the local gas density encourage the birth of new stars, which in turn causes the propagation of fresh waves of compression. The bright blue colour of the spiral arms is due to the presence of young stars.

These processes also have a tendency to progressively change the structure of the disc. If we start with a uniform disc, we will first form spiral arms due to the density waves. But after the galaxy has rotated a few times the arms will tend to be replaced in the central regions by a more elongated structure resembling a bar. Such bars may contain as much as 40 per cent of the visible matter in a galaxy. Thus, we have an overall picture of how barred spirals form, and a prediction that they should occur at a later stage in the evolution of spiral galaxies. Unfortunately, it is difficult to test these models directly even though most astronomers believe that the general scenario is correct.

Though the magnificence of such a galaxy is due to the unfolding of the spiral arms, these arms only represent a very small part of the galaxy; in fact, at most only a few per cent of the mass of the galaxy is contained in them! The arms are visible so prominently because they are the active regions of the galaxies, in which the majority of stars form and we find the brightest ones.

In reality, a spiral galaxy is made of at least four distinct components. First is the central region of a spiral galaxy with a bulgy appearance which, when viewed in isolation, looks like a miniature elliptical galaxy. Second is the flattened disc-like structure in which the spiral arms reside. The spiral arms contain not only fully formed stars but also large quantities of gas and dust with differing chemical composition. This (third) component may be called the 'interstellar medium' and plays a vital role as the womb for new stars. Finally, one often finds that even the disc-like structure is embedded in a very tenuous, extended, spherical region ('halo') which contains aggregates of stars in the form of 'globular clusters'. These globular clusters are collections of about a million stars in approximately spherical shape distributed all around the galaxy.

There are vital differences in the composition of the disc and the bulge. The bulge contains comparatively older stars than the disc. Most of the action, including current star formation, is in the disc, while the bulge is comparatively inert. In fact, detailed studies have shown that the bulge and disc, components of a spiral galaxy must have formed in two different stages. The bulge formed rapidly in the beginning, followed by a slow and prolonged star formation in the disc. This model, for example, explains the differences in the abundance of elements in the bulge and in the disc. Remember that heavy elements are synthesized inside the first generation of stars and distributed in the galaxy via supernova explosions; the second generation of stars forming from this debris will, therefore, have more of the heavy elements. Having formed first, the bulge should contain comparatively fewer heavy elements than the disc. This is indeed the case.

(b) Elliptical galaxies

At the other extreme of the galaxy family are the elliptical galaxies, which look somewhat less impressive than the spirals. The photograph of an elliptical galaxy might suggest that such an object is symmetrical about an axis; i.e., if we rotate the galaxy about an axis its shape will not change. But a little thought will show that we are actually falling prey to an optical illusion. Whatever the real shape of the galaxy, we get to see only its projection on the sky. Suppose, for example, that the actual shape of an object is an elliptical *disc*. If we happen to see it face on, it will look like an elliptical galaxy to us and we may erroneously conclude that there is just as much matter in the line-of-sight as perpendicular to it. In other words, it is quite possible that the real shape of elliptical galaxies is far more complex than suggested by their two dimensional photographs. A shape like the one shown in figure 4.5 will appear to be an ellipse when projected on the sky along any of the three axes. But since the size of the object is different along the three different axes, these ellipses will also be different. Such a body is called a 'triaxial ellipsoid'. If the elliptical galaxies have such a shape and the photographs of the sky only give a two dimensional projection along some arbitrary direction (not necessarily along the axis), it is quite difficult to divine the true shape from the projection.

Originally, astronomers thought that elliptical galaxies have the shape shown in figure 4.5 but with $c = a$, because it was known that this is the shape acquired by a rotating self-gravitating gas. In this case, the elliptical galaxy would be symmetric about the long axis. However, with very careful study of the distribution of light in an elliptical galaxy, coupled to the information of the motion of stars obtained from Doppler shifts, they have now found that

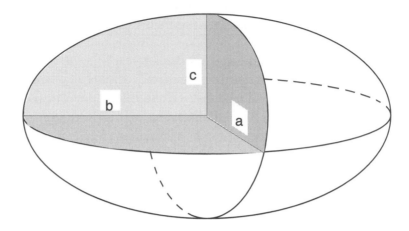

Figure 4.5   The body shown in this figure ('ellipsoid') will look like an ellipse when viewed along any of the three axis. But if the sides *a*, *b* and *c* are not equal it will look different when viewed along each of the axes and will not have any symmetry. Elliptical galaxies have shapes like the one shown above.

most elliptical galaxies are quite asymmetric. That is, their sizes along three perpendicular directions are unequal and the shape is a triaxial ellipsoid.

The ellipticals differ from the spirals in many other ways. By and large, ellipticals do not show signs of much activity compared to spirals. They contain relatively old, colder, stars of the red giant type which are rich in heavy elements. The absence of young stars and the presence of heavy elements in old stars suggest that the combined processes of star formation, supernova explosion and chemical enrichment have occured much more rapidly in the ellipticals than in the spirals. The characteristic timescale seems to be few hundred million years, which is quite short by cosmic standards.

The elliptical galaxies do contain a large amount of neutral hydrogen, which emits a characteristic radiation at a wavelength of 21 cm detectable by radio telescopes. The ellipticals also contain large quantities of dust. In these two aspects they are somewhat similar to spirals but only superficially.

Another remarkable feature of the ellipticals is their relative homogeneity as a family. For example, astronomers have found that an elliptical galaxy can be characterized essentially by its luminosity. All other characteristics (like, for example, the mean speed of the stars) can then be determined from the luminosity. This uniformity is remarkable, since elliptical galaxies exist with masses varying anywhere from a hundred million solar masses to ten million million solar masses.

Having settled the appearance of different galaxies, the astronomer has to consider their evolution. Here the situation is far more complicated than in the case of stars. As far as the stars are concerned, we have direct observational evidence relating to all phases in the evolution of a star, starting from its birth to its death, and hence the theoretical modelling is also somewhat easy. In the case of galaxies, we are only beginning to have a consistent picture of how they were born and how they evolve.

It is believed that galaxies formed due to the gravitational contraction of large, diffuse, gas clouds which originally had a small amount of rotation about an axis. As the gas cloud contracts, individual lumps in it will also contract locally, forming stars. If the stars form fast enough, one obtains a somewhat ellipsoidal distribution of stars held together by self-gravity – which could account for elliptical galaxies. On the other hand, if the gas cloud collapses to a plane perpendicular to the axis of rotation, even as the stars are forming, then a more complicated configuration can arise. As the gravitational attraction pulls the cloud together it will begin to spin faster. The first generation of stars, including globular clusters, will be formed around this time. After about 100 million years from the time such a galaxy has assumed its basic shape, the gas falling to the plane from the two directions collides and eventually makes up the disc. As time passes, the next generation of stars form in the disc and the galaxy assumes a well defined disc-like shape. The density wave mechanism described above leads to enhanced star formation along the spiral arms, thereby leading to the structures we see today. (See figure 4.6.)

The formation and evolution of stars through different generations will necessarily imply an evolution for the galaxy. This transformation, however, is somewhat inefficient because the remnants of stellar evolution (white dwarfs, neutron stars and black holes) lock up a sizable fraction of matter in a form which is not reusable for star formation. Every new generation of stars has to form out of the fraction of mass which was ejected out of the previous generation of stars. Thus, without any source for new gas, the rate of star formation in galaxies must decrease with time. In addition, we must remember that low mass stars live much longer than the high mass ones; so, as time progresses, low mass stars will accumulate in a galaxy. Because of these two facts, the properties of galaxies, particularly the luminosity and colour, will change with time.

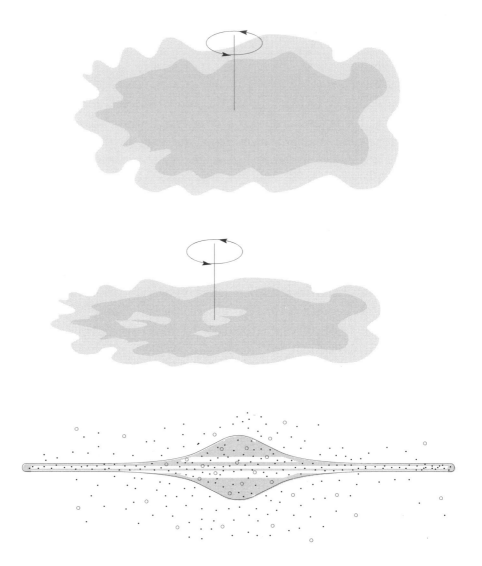

Figure 4.6 This schematic illustration summarizes the current belief regarding the formation of disc galaxies. The top frame shows a diffuse gas cloud rotating about some axis. As it contracts under self-gravity, it will start spinning faster and the gas will settle down on a plane perpendicular to the rotating axis. The individual lumps within the gas will also contract and form the first generation of stars. Later on, the disc forms, and the second generation of stars is produced from the disc material.

A more interesting and energetic form of evolution takes place due to gravitational dynamics. We saw one example of this in the formation of barred spirals due to gravitational instability of the normal spiral. Such gravitational instabilities can be accelerated due to close encounters between galaxies. In fact, many of the shapeless peculiar galaxies can be explained as due to disruption or merging arising out of galaxy encounters. A more extreme hypothesis along the same line is that elliptical galaxies themselves arise as a result of collision between two disc galaxies. In a collision between two spirals, the individual stars are unlikely to hit each other because the typical size of the star is much smaller than the typical distance between them. But the gas contained in the disc will collide directly. It is difficult to predict the outcome of such complex dynamics, but it seems likely that compression of the gas will enhance star formation. Though it is difficult to believe that *all* ellipticals have formed this way, it is possible that some ellipticals have such an origin.

The formation of galaxies is one of the key research areas today and we shall say more about it in chapter 6.

4.7 The Milky Way – our home

The galaxy which has been studied most extensively, of course, is our own. The Milky Way happens to be a typical spiral galaxy with an overall shape which is similar to that of the majority of spiral galaxies; viz., that of a disc with a bulge at the centre and a very tenuous halo surrounding it. The disc itself is nearly 30 kpc across and is fairly flat; the thickness in the outer regions is only about 300 pc. The central bulge has a diameter of about 6 kpc and a thickness of 1 kpc.

Like other galaxies, the Milky Way has stars, dust and gas. They occur in different proportions in each of the three components of the galaxy, namely in the disc, bulge and halo. In the disc, for example, nearly 10 per cent of the matter is in the form of an interstellar medium of gas and dust, while 90 per cent is in the form of stars. In the bulge, the interstellar matter is only about 1 per cent, while 99 per cent is made of stars. The halo has a negligible amount of matter compared to the disc or the bulge.

The dust, composed of carbon, silicates and other material, causes absorption of light by different amounts in different directions, thereby giving rise to the appearance of 'holes' in the stellar distribution of the Milky Way. For this reason, it is not easy to determine the shape of our galaxy 'from the inside'. It is the observations in the radio band which allow one to penetrate the dust and gas and confirm that our galaxy *is* a spiral and – in fact – contains four distinct spiral arms (see figure 4.7). These arms are named after prominent constellations in the sky. For example, the one farthest from the centre lies around the constellation of Perseus. At the centre of the galaxy, as in all spirals, there is a concentrated core of stars forming a bulge. Viewed from the Earth, the central bulge lies in the direction of Saggitarius.

The three components of our galaxy (disc, bulge and halo) also contain different kinds of stars, on the average. The stars in the disc are relatively young and called Population I stars. They orbit around the centre of the galaxy very much like the planets going around the Sun. Their ages range from a few million years up to 12 billion years. In fact, the youngest of the stars in the disc are young even when compared to the solar system (which is about 4.5 billion years old). Star formation is an ongoing activity in the disc.

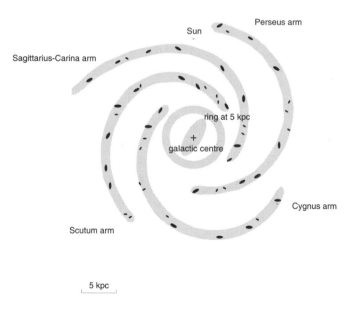

Figure 4.7   The spiral arms of the Milky Way galaxy can be delineated by the emission of radiation from hydrogen gas at a wavelength of 21 cm. Such studies show that there are four distinct spiral arms to the Milky Way.

The stars in the bulge and halo (called Population II stars) are considerably older and could have formed about 13 billion years ago. Further, the halo stars occur in concentrated spherical clusters known as globular clusters. The stars in these globular clusters are probably the oldest objects in our galaxy. They are relatively poor in elements heavier than hydrogen and helium.

The average separation between stars in the Milky Way is a few parsecs. The space between the stars is hardly empty, and is occupied by diffuse matter whose total mass is about 10 per cent of that in stars. It is this 'interstellar medium' from which new stars are born. The nature of the medium varies from place to place in the galaxy and requires different techniques for its observations. Studies show that there are at least two components in the interstellar medium. The first one is made of dense clouds of gas, containing tens of thousands of particles per cubic centimeter, at a temperature of about 100 K. Since this temperature is fairly low by astronomical standards, this is known as the 'cold' component. The second component is a dilute, tenuous, medium in which the above clouds are embedded. This medium is principally made of hydrogen, though it contains a small fraction of different chemical species, including molecules as complex as HC_9N. The interstellar medium also contains appreciable quantities of tiny solid particles called interstellar dust.

The clouds of gas in the Milky Way appear very differently in photographs. For example, the 'coal-sack' nebula seen from the southern hemisphere appears as an extended dark region, since it is opaque and blocks the light of the stars behind it. There are other nebulae which become somewhat more visible when they glow under the impact of stellar radiation arising from very hot stars embedded in their midst. Some of these nebulae are active regions of star formation, and provide direct evidence regarding the birth and infancy of stars to astronomers.

One such region, which has been studied extensively, is the Orion nebula located just below the belt of Orion. (On a clear night you can see it with the naked eye; with a pair of binoculars it is a spectacular sight.) This region turns out to host several very young stars – some of the stars could be less than 10 million years old. Near this nebula there are several regions of interstellar clouds, which are detected by radio telescopes. They contain groups of stars, with the oldest being nearly 8 million years old and the youngest less than a million years old (see figure 4.8). The 'young stars' are not yet hot enough to emit visible light, but they shine brightly in the infrared band. The study of such nebulae shows how stars form out of interstellar clouds and how – over a period of time – they are dispersed all over the galaxy.

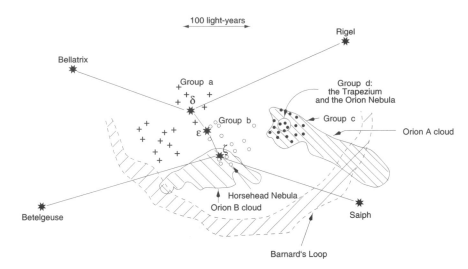

Figure 4.8   In the constellation Orion there exist several interstellar clouds which are detected by radio observations. They foster active regions of star birth. In the figure, group *a* stars are the oldest (8 million years) followed by group *b* (5 million years), group *c* (3 million years) and group *d*, which are the youngest, having an age of less than a million years.

This entire gamut of matter in our galaxy – stars, interstellar clouds, gas and dust – is in a state of motion. By and large, they all rotate in the disc of the galaxy around the centre. The rotation is not uniform; stars which are farther out from the centre take a longer time to complete an orbit. (The Sun, for example, takes about 200 million years to complete one orbit.) The rotation of stars and gas clouds in a disc galaxy provides valuable information to astronomers. To see its importance, let us ask the question: how much mass is contained in a galaxy within some distance from the centre? One way to answer this question would be to carefully count all the stars which are located within this distance. Knowing the relationship between the luminosity and mass of stars of different kinds, one can calculate the total amount of mass which is in the form of stars in any galaxy. One could also do the same as regards the interstellar medium, though it contains a negligible amount of mass compared to the stars. The total mass of the galaxy which we arrive at through the above process is called the 'visible mass', since it was detected through the electromagnetic radiation which it emits. The question, of course, is whether there is more to the galaxy than meets the eye. Is there matter present in the galaxy which is not emitting electromagnetic radiation? In other words, is there matter in the galaxy which is dark? The rotation of stars and gas clouds provide a way of answering the above question. A star or a gas cloud at the edge of a galaxy will feel the gravitational force of *all* the matter which is inside, rather than just the visible matter. By measuring the speed of rotation of gas clouds and their distance from the centre of the galaxy, one can actually compute the total amount of mass which is within that distance. It turns out that galaxies contain nearly 10 times more mass than that determined by counting the stars; we shall say more about this in chapter 5.

The above discussion should indicate that we understand far less about the galaxies than, say, about the stars. For example, we do not have anything analogous to the H-R diagram for the galaxies, nor do we have a model for the formation and evolution of galaxies. These are the issues we will be concerned with in the later chapters.

4.8   Groups and clusters of galaxies

Telescopes reveal that there are more than a billion galaxies in the visible universe. One may wonder how these galaxies are distributed in space. Are they distributed uniformly or is there any specific pattern in their arrangement? It turns out that they are *not* distributed uniformly in space. They show a dominant tendency to cluster among themselves, though some fraction of galaxies are seen in isolation. Loosely speaking, one may think of clusters of galaxies as the next higher level in the hierarchy.

A galaxy cluster may contain anything from about 10 galaxies to nearly 1000 galaxies. By convention, collections of about 10 to 40 galaxies are called 'groups' and aggregates containing 40 to 1000 members are called 'clusters'. We shall use these terms interchangeably.

Our own galaxy, the Milky Way, is a member of a group called the Local Group, which has a size of about 1.3 Mpc. To understand the level of clustering involved in the Local Group, one may note the following fact. A region of size 1.3 Mpc around our galaxy contains nearly 30 or so galaxies; the region between 2.4 Mpc and 15 Mpc around us contains several thousand galaxies. However, there are no galaxies at all in the shell between 1.3 Mpc and 2.4 Mpc! This clearly shows that the Local Group can be thought of as an island-like region in the local neighbourhood. The entire universe is punctuated by islands of this kind.

If we take a walk around the Local Group, we will see a wide variety of structures. The major members of the Local Group are two giant spiral galaxies, the Milky Way and Andromeda. These two are separated by nearly 700 kpc and are gravitationally bound to each other. In addition, there are two other (somewhat average-looking) spiral galaxies, one elliptical galaxy, half a dozen or so small irregular galaxies and a dozen dwarf galaxies (each with only about a million stars) in our Local Group. By and large, the smaller galaxies are clustered around either the Milky Way or Andromeda (see figure 4.9). This means that the Local Group does not have a central condensation and, in fact, there is very little in the centre. The entire structure is held together by the gravitational interaction of its components.

Among the satellite galaxies associated with the Milky Way, the two closest companions are the Large and Small Magellanic Clouds (LMC and SMC, for

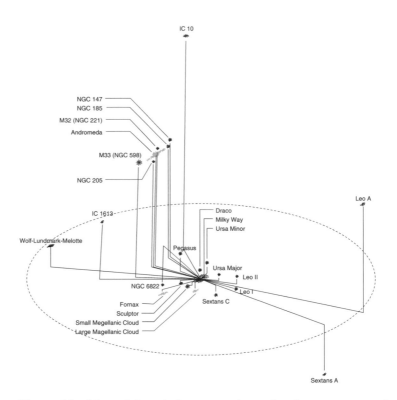

Figure 4.9   Most of the galaxies are members of a cluster or group of galaxies. Our galaxy, the Milky Way, for example, belongs to a group of 30 or so galaxies, all gravitationally bound to each other. The Milky Way and another large spiral galaxy, called Andromeda, are the two most prominent members of this Local Group. Most of the other galaxies are smaller in size and are either clustered around the Milky Way or around Andromeda.

short). These are relatively massive, gas rich, irregular galaxies which show evidence for recent star formation. They are approximately 60 kpc away from the the Milky Way. The LMC has a central bar-like structure and highly truncated spiral arms, and contains regions of extensive star formation. In fact, the star forming regions in the LMC are intrinsically much brighter than the Orion nebula we discussed above. The two Magellanic Clouds appear to be connected by a 'bridge' of neutral hydrogen gas. This is thought to be possibly due to the interaction between the the Milky Way and the Magellanic Clouds.

The dwarf elliptical galaxies, which account for nearly half of the members of the Local Group, are more prosaic compared to the Magellanic Clouds. They have very little gas and are made of fairly red stars, concentrated towards the centre of the galaxy. Because of this feature, dwarf galaxies are hard to detect, and we are not even sure whether we have detected all the dwarfs which belong to the Local Group.

The Andromeda galaxy is, of course, the largest and most massive member of the Local Group and resembles the the Milky Way in several respects. The disc of Andromeda has two spiral arms marked out by dark dust lanes, and several young star clusters. Like other spirals, Andromeda also shows a bulge and disc and distinct stellar populations. The satellites to Andromeda include two examples of (more or less) normal elliptical galaxies of the Local Group. They are somewhat small compared to a typical elliptical, but their structure is clearly different from that of the dwarfs.

The Local Group is only a prototype of the organization of galaxies in the universe. Several such groups of galaxies have been seen in the universe. Typically, each group has a dozen or so galaxies within a volume of about one cubic megaparsec. This number density is at least ten times larger than the background density of galaxies in the universe. The nearest group of galaxies to the Local Group is called the 'Sculptor Group', which is about 2.4 Mpc away. The Sculptor Group contains half a dozen or so bright spiral galaxies distributed within a region of size 1 Mpc (see figure 4.10). Richer clusters of galaxies contain over 100 galaxies clustered together in a region of a few megaparsec in size. Because of the large density of so many galaxies, clusters can be seen even at great distances, and probably constitute the largest entities in the universe which are bound by self-gravity.

The nearest rich cluster is a large, somewhat haphazard concentration of galaxies in the direction of the constellation Virgo, at a distance of about 20 Mpc. Like many other large clusters, its shape is somewhat irregular, and its

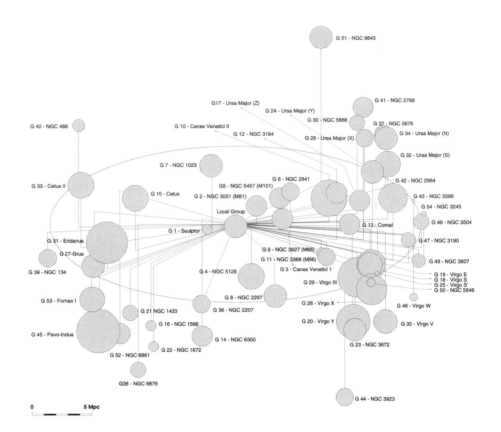

Figure 4.10   Within a radius of nearly 15 Mpc from our galaxy, one finds over 50 groups or clusters of galaxies. This figure shows these groups with our Local Group in the centre. Each group is shown by a figure whose radius indicates the approximate size of the group. It is clear that there are a greater number of groups on the right hand half of the figure. This is because our Local Group lies at the edge of a large local supercluster of galaxies which is on the right hand side.

brightest galaxies are ellipticals which lie near its centre. Another comparatively large cluster is the Coma cluster, located at a distance of about 100 Mpc. Unlike the Virgo cluster, Coma is a very rich cluster with nearly spherical shape and more than 1000 bright galaxies. The central part of the Coma cluster contains a large number of ellipticals and S0 galaxies without many spirals. It is generally believed that the population of ellipticals in a rich cluster has been influenced by evolutionary processes like mergers, etc.

We saw earlier that the region between the stars in a galaxy is not empty and is populated by gas and dust. What about regions between galaxies? It turns out that they are also not empty. In a cluster, for example, there exist large quantities of gas in between the galaxies. This gas is very hot (nearly 10 to 100 million degrees Kelvin) and emits copious amounts of X-rays. Even when a galaxy is not part of a large cluster, there is evidence to suggest that it is embedded in a tenuous intergalactic medium of ionized hydrogen and helium gas.

Our tour of the universe has now taken us from stars to galaxies and to clusters of galaxies. At each level, the objects are bound by gravitational interaction. Just as the planets maintain their orbits in the solar system with their motion balancing against their gravitational attraction, it is the motion of constituent objects which maintains the larger structures. In a galaxy, the stars move under their mutual gravitational attraction; in a cluster, the galaxies move under the action of their self-gravity. We may be led to ask the natural question. Are there still larger scale motions present in the universe? The answer happens to be 'yes'. In fact the entire universe is participating in a spectacular form of motion which leads to a picture of an expanding universe. We shall look at the overall expansion of the universe in the next chapter.

5

The expanding universe

5.1 Expansion of the universe

The cosmic tour which we undertook in the last chapter familiarized us with
the various constituents of the universe from the stars to clusters of galaxies.
We saw that the largest clusters have sizes of a few megaparsec and are separated typically by a few tens of megaparsec. When viewed at still larger scales,
the universe appears to be quite uniform. For example, if we divide the universe into cubical regions, with a side of 100 Mpc, then each of these cubical
boxes will contain roughly the same number of galaxies, clusters, etc. distributed in a similar manner. We can say that the universe is homogeneous when
viewed at scales of 100 Mpc or larger. The situation is similar to one's percep-
tion of the coastline of a country: when seen at close quarters, the coastline is
quite ragged, but if we view it from an airplane, it appears to be smooth. The
universe has an inhomogeneous distribution of matter at small scales, but when
averaged over large scales, it appears to be quite smooth. By taking into account
all the galaxies, clusters, etc. which are inside a sufficiently large cubical box,
one can arrive at a mean density of matter in the universe. This density turns
out to be about 10^{-30} gm cm^{-3}.

The matter inside any one of our cubical boxes is affected by various forces.
From our discussion in chapter 2 we know that the only two forces which can
exert influence over a large range are electromagnetism and gravity. Of these
two, electromagnetism can affect only electrically charged particles. But since
matter on the large scale is mostly neutral, we can ignore any electromagnetic
influence in the overall dynamics of the universe. We thus arrive at the conclusion that the large-scale dynamics of our universe is entirely governed by the
gravitational force.

What kind of gravitational force is exerted by a uniform distribution of
matter in the universe? We cannot, unfortunately, answer such a question
using the Newtonian theory of gravity. When the material content of a large

region of the universe needs to be taken into account, the gravitational fields can be rather large, and it is necessary to use the correct theory of gravity – namely, Einstein's theory of gravity – to attack such a question. Fortunately, Einstein's theory provides an extremely simple description of a universe which contains a uniform density of matter: it predicts that such a universe will be in a state of expansion, with the distance between cosmic objects (like galaxies) increasing with time. Mathematically, such a situation is described by using a quantity called the 'expansion factor'. The expansion factor determines the overall scale in the universe. As time goes on, this expansion factor increases, thereby describing the increase in physical length scales.

This model suggests that all distant galaxies must be moving away from our galaxy due to cosmic expansion. This is indeed true. As the astronomer looks deeper into space, (s)he sees the galaxies located at farther and farther distances to be moving away with larger and larger speeds. To understand the features of our universe, we have to first understand the nature of this expansion.

If all the other galaxies are flying away from us, it is rather tempting to think that we are at the centre of the universe. But in reality, there is no such thing as the centre of the universe; the above feeling is illusionary. In fact, any other observer – located at any other galaxy – will also see the same effect. To understand this, it is helpful to consider a simpler example. Imagine a very long stretch of road with cars moving in one direction. Assume that, at some instant of time, the distance between any two consecutive cars is 100 meters and that you are sitting in some chosen car, say, A (see figure 5.1). Let the first car (B) in front of you move at 10 km hr^{-1} with respect to you, the second one in front (C) at 20 km hr^{-1}, the third (D) at 30 km hr^{-1}, etc. It will appear to you as though all the cars are moving away from you with speeds proportional to the distance from you. But consider a man sitting in the car (B) just in front of you. How will he perceive the car just in front of him? Since he is moving at 10 km hr^{-1} with respect to you, and the second car (C) in front of you is moving at 20 km hr^{-1} with respect to you, the driver of the car (B) will think that the car (C) is moving only at 10 km hr^{-1} with respect to him. By the same reasoning he will conclude that car (D) in front of him is moving at 20 km hr^{-1}, etc. In other words, he will come to the same conclusion regarding the motion of vehicles as you did! Clearly, there is nothing special about your position. All the cars in this chain enjoy equal status.

The motion of galaxies in the universe is quite similar. Every observer, irrespective of which galaxy he is located in, will see all the galaxies moving away from him with a speed progressively increasing with distance. This cur-

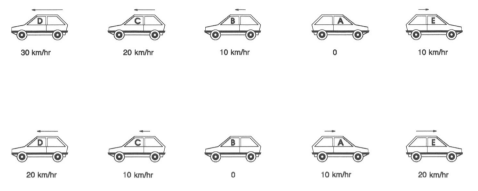

Figure 5.1 A sequence of cars is moving along a long stretch of road with the speeds as shown in the figure. The top part of the figure shows the situation as viewed from the car A. The driver of A sees the four other cars as moving with speeds which increase in direct proportion to their distance. The driver of this car might think that his position is 'special' and that all the cars are moving away from him. But the driver of car B will see the car C as moving at 10 km hr^{-1}; this is because car B is moving at 10 km hr^{-1} with respect to car A and car C is moving at 20 km hr^{-1} with respect to car A. Clearly, the driver of car B will also think that all the cars are moving away from his car.

ious motion of the galaxies is what really constitutes the 'expansion of the universe'. After all, the word 'universe' is merely a convenient terminology for describing the matter which we see around us. It is the movement of the parts of the universe which is indicated when one talks of the expansion of the universe.

Lack of understanding of this fact can easily lead to several confusing questions. The first confusion is: 'If the whole universe is expanding, what is it expanding *into*?' Notice that the manner in which we have described the expansion of the universe makes this question irrelevant. In fact, one need not even talk about 'expansion'; all we know from observations is that distant galaxies are moving away from us with speeds increasing in direct proportion with the distance. Surely, in an infinite universe such a motion can take place without leading to any contradiction. If we understand the term 'expansion of the universe' as representing 'motion of distant galaxies away from us' then such confusions will not arise.

The second confusion is the following: 'If everything in the universe is expanding (galaxies, stars, Earth, meter scales, etc.) then how will we ever know that there is expansion?' This is simply incorrect. Everything in the universe is *not* expanding. All that is happening is an increase in the distances between galaxies which are sufficiently far away from each other. Remember that the expanding model for the universe was based on Einstein's equations for gravity as applied to a *uniform* distribution of matter. When the deviations from uniformity become important, we have to take into account the self-gravity of these structures and cannot use the results obtained for a uniform distribution of matter. The stars or the Earth or the meter scales which we use for measurements are *not* expanding. Once again this confusion arises because of the loose interpretation of the term 'expansion of the universe'.

Having settled the terminology, let us get back to the physics. How fast is the universe expanding? Or, rather, how fast are the galaxies receding from us? This can be determined from the study of the spectrum of light emitted by the galaxies. If the galaxies are receding from us, then the lines of the spectrum will be shifted to the redder side (that is, they will be 'redshifted") due to the Doppler effect (see chapter 2, section 2.4). By measuring the amount of Doppler shift, one can find out the speed of recession of the galaxy. If we also know the distance to the galaxy, then we can figure out how the speed of recession changes with distance. Such studies show that the speed of recession is proportional to the distance; i.e., a galaxy twice as far away will be moving away from us with twice the speed. The rate of such an expansion is usually denoted by a quantity called the 'Hubble Constant', which has a value between 50 and 100 km s^{-1} Mpc^{-1}. To indicate this uncertainty in the observed value of the Hubble constant, cosmologists usually write it as $H = 100h$ km s^{-1} Mpc^{-1}, where h is a fraction having a value between $1/2$ and 1. (If $h = 1/2$, then $H = 100 \times 1/2$ km s^{-1} Mpc^{-1} = 50 km s^{-1} Mpc^{-1}, etc.) This uncertainty has persisted now for decades, and arises because distances to galaxies are not known with a high amount of accuracy. Several observational astronomers are currently working on different methods of measuring the Hubble constant to greater accuracy. This is particularly important because the value of the Hubble constant is intimately related to the age of the universe; smaller values of the Hubble constant imply larger age and vice versa. Taking its value to be 50 km s^{-1} per Mpc, one can conclude that the universe will double its size in about 18 billion years.

The fact that the galaxies have a systematic large-scale motion forms the cornerstone of modern cosmology and leads to several important consequences. To begin with, it shows that the universe is continuously changing and evolving. Of course, these changes take place over billions of years and hence are not of practical consequence in everyday life. Nevertheless, it shows that the universe is dynamic and not static. Because of this dynamism, the physical conditions of the universe would have been very different in the past – say, billions of years ago – compared to the present.

Further, since structures in the universe are moving away from each other today, it follows that they would have been closer together in the past. The universe would have been denser in the past with the density increasing as we go farther and farther back in time. Such an increase in the density would have also made the particles in the universe move faster and have higher temperature. We are thus led to the conclusion that the early universe would have been considerably denser and hotter.

This leads to another startling result. Most of the matter we come across in everyday life will change its form at high temperatures. For example, we saw in chapter 2 that normal matter in the form of neutral atoms cannot exist at temperatures higher than, say, a million degrees. The electrons in the atoms would have been separated from the nuclei at such high temperatures, and the matter would exist only in the form of a plasma. It follows that, at sufficiently early epochs in the evolution of the universe, all matter must have existed in such a plasma state. We shall explore the physics of this phase of the universe in the next few sections.

What happens when we go farther and farther back in time? According to Einstein's theory, the size of the universe would have been zero at some finite time in the past, nearly 15 billion years ago. All matter would have been compressed to a point, and the density and temperature of the universe would have been infinite. Such a moment is called a 'singularity' in mathematics and the 'big bang" in popular literature! You must remember that this is just a theoretical speculation arising out of Einstein's theory. There are reasons to believe that Einstein's theory breaks down when the size of the universe is very small, and has to be replaced by a better theory (though nobody knows what the better theory is; we shall say more about it in section 5.5). Thus the 'big bang' is just a convenient way of talking about a moment at which Einstein's theory breaks down, and we usually set the clock to zero at this moment. Since it *is* convenient, we will continue to use expressions like 'when the universe was one second old', but you must remember that it is only a manner of speaking.

There is a very different aspect to the expansion of the universe which needs mentioning at this stage. In the above discussion, we started from the correct law of gravitation (namely, Einstein's theory) and argued that this theory demands an expanding universe. Actually, there is a very different way of thinking about the distribution of matter in the universe which goes under the name 'Olber's paradox'. Let us suppose for a moment that the universe is not expanding but contains a uniform distribution of luminous objects, say, galaxies. At first sight, such an infinite universe – with luminous objects distributed all over space – seems to agree with our experience. It is therefore remarkable that one can actually prove that our universe is *not* made that way. This argument runs as follows.

Consider any given point in the universe, say, on Earth. Let us consider an imaginary spherical shell of radius R and thickness X, centred on Earth. What will be the intensity of light on Earth due to all the objects within this shell? The intensity of light from any luminous body falls as the square of the distance from that body. (If you go twice as far from the body, the intensity will decrease by a factor of 4, etc.) In a universe which has uniform distribution of matter, the amount of matter contained in a shell, at a given distance from any particular point, increases as the square of the distance. Since these two effects nicely cancel each other, the amount of intensity on Earth due to any given shell of matter is independent of the distance R to that shell. Objects located in the shells which are farther from Earth contribute less intensity but there are more of them. So, each shell of matter ends up contributing the same amount of intensity at any given point. Since there are infinite number of such shells, the total intensity should be infinitely large! This conclusion gets slightly modified when we take into account the fact that each luminous object has some definite size. You can convince yourself that, in such a case, any point in the sky will be as bright as the luminous surface of the body – i.e., any point on the sky should be as bright as the surface of the Sun.

So we reach the surprising conclusion that in an infinite universe with uniform distribution of luminous objects, the night sky should be as bright as the surface of the Sun. This conclusion is clearly false, and shows that the simplest theoretical model one can think of for our universe is not viable.

There are, of course, several ways of getting out of this paradox. Firstly, if the universe has a finite age, then the stars could have been shining only for a finite period of time, and there might not have been enough time for the light from a given star (or galaxy) to travel an infinite distance. Secondly, if the universe is expanding, then the above considerations do not apply. According to the current cosmological models, the universe is expanding *and* stars have a finite age; so we need not worry about Olber's paradox in our models. However, it serves as a powerful example of how much can be achieved by 'pure thought'. The darkness of the night sky is a significant clue to the behaviour of our universe.

5.2   The universe has a past!

We saw in the last section that the physical conditions in our universe change with time because of the expansion. By tracing the expansion of the universe back in time, one can try to understand the physical conditions that would have existed in the past. Such an exercise leads to a very fascinating picture of the early universe.

When the universe expands, the density of matter (and radiation) decreases. As we go into the past, on the other hand, the universe becomes more and more dense. Today, the matter density in the universe (taking into account both visible and dark matter) is about 5×10^{-30} gm cm^{-3}. When the universe was 10 times smaller, the matter density would have been 1000 times higher because the volume of the universe varies as the cube of its (linear) size. However, something curious happens to the energy of the *radiation* which exists in the universe. When the universe was 10 times smaller, the radiation energy would have been 10 000 times higher (rather than just 1000 times higher). To understand the reason, it is best to think of radiation as made of a large number of photons. We saw in chapter 2 that the energy of a given photon is inversely proportional to its wavelength. As the universe expands, the wavelength of each photon gets stretched. In the past, when the universe was 10 times smaller, the wavelength of each photon would have been 10 times smaller; so the energy of each photon would have been 10 times larger. The number density of photons would also have been higher by a factor of 1000. (This factor, of course, is the same for matter density and photon number density.) Together, the energy density contributed by the photons would have been 10 000 times higher when the universe was 10 times smaller.

This leads to an interesting conclusion. Since the energy contributed by radiation increases faster than the energy contributed by matter (as we go into the past), it is clear that at some sufficiently early phase, radiation energy will dominate over matter. To figure out when this will happen, it is necessary to know how much of the energy in the universe is in the form of radiation at present. We have to take an inventory of all the radiation present in the universe today.

The electromagnetic radiation in the present universe can be divided broadly into two parts. The first part is the radiation which arises from specific sources in the sky. Radiation from the stars, galaxies, etc. falls in this category. It could be in different wave-bands depending on the characteristics of the source which is emitting it. For example, a star like the Sun will emit a large amount of optical or infrared radiation, while hot gas in a cluster of galaxies will emit dominantly in X-rays. Whatever may be the wave-band, this kind of radiation can be specifically identified with sources in the sky. There is a second category of radiation which is present in our universe, called 'background radiation'. This is the radiation which exists all around us in a more or less uniform manner, and cannot be identified with any specific source in the sky. Astronomers have found that the universe does contain background radiation in several wave-bands like microwaves, X-rays and gamma rays.

Comparing the amount of energy present in the first category of radiation (which arises from specific sources like the stars, etc.) with the second, it is found that most of the energy is indeed in the second category; that is, most of the energy is distributed as a uniform background in the sky. What is more, the dominant part of this background radiation has the characteristics of a thermal spectrum. We saw in chapter 2 that material bodies kept at some temperature will emit radiation with a characteristic spectrum (see figure 2.14) determined by the temperature alone. The background radiation in the universe has this particular shape, with a characteristic temperature of about 2.7 K. It is as though the entire universe is enclosed by a box kept at this temperature. The thermal radiation at a temperature of 2.7 K will peak in the microwave region of the electromagnetic spectrum. Hence this radiation is usually called 'microwave background radiation'.

The energy density of the microwave radiation can be computed from its temperature. It turns out that this energy density (at present) is nearly 10 000 times smaller than the energy density of matter. But since radiation energy density increases faster than matter density, it will have caught up with matter density sometime in the past. When the universe was smaller by a factor of 10, the matter density would have been higher by a factor of 1000, while the radiation density would have been higher by a factor of 10 000. So when the universe was 10 times smaller, radiation had gained a factor of 10 in density compared to matter. Since radiation density today is down by a factor of 10 000, it will have caught up with matter when the universe was 10 000 times smaller. At still earlier epochs the universe will have been dominated by radiation and not by matter (see figure 5.2).

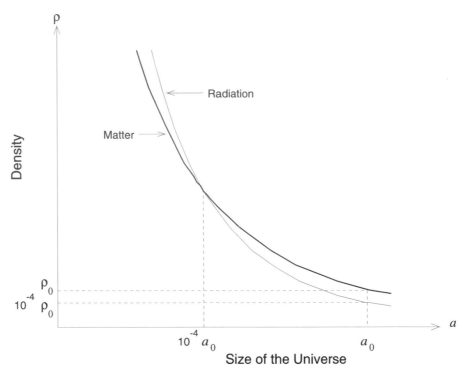

Figure 5.2 The radiation energy density (shown by the thin line) grows faster than the matter energy density (thick line) as we go into the past. Today, when the size of the universe is a_0, say, the matter density is ρ_0 and the radiation density is about ten thousand times smaller. As we go into the past, the thin line rises faster and catches up with the thick one when the size of the universe was ten thousand times smaller. At still earlier epochs, the universe was dominated by radiation rather than matter.

Cosmologists use a convenient terminology based on the concept of redshift to speak about the early universe. We know that, as the universe expands, the wavelengths of photons are stretched, thereby introducing a redshift. When the universe expanded by a factor of 100, say, a wavelength λ would have become 100λ. The ratio between the increase in the wavelength ($100\lambda - \lambda = 99\lambda$) to the original wavelength (λ) is called the 'redshift z' (which is 99 in this case). This example shows that a redshift of z corresponds to an epoch when the universe was smaller by a factor of $(1 + z)$. So instead of saying 'when the universe was smaller by a factor of 200... ' we might as well say 'when the redshift was 199... '. For most practical purposes, when we are talking about the early universe, z will be sufficiently high for us to ignore the number 1 in

$(1 + z)$. With this convention, we will often say that the radiation density was equal to matter density at a redshift of $z = 10\,000$. This is just a way of talking about various epochs in the early universe, but is extremely convenient.

As the density of the radiation increases, its temperature also goes up. When the universe was 1000 times smaller, it also would have been 1000 times hotter, having a temperature of about $1000 \times 2.7 = 2700$ K. As we go further into the past, the size of the universe will become smaller and the temperature and density in the universe will increase more. According to theoretical calculations, the temperature and density of the universe would have reached infinite values at some finite time in the past, nearly 15 billion years ago. But, as we said earlier, one cannot have faith in a theoretical calculation which predicts infinite density and temperature, and it is very likely that the theoretical models on which these calculations are based will break down well before such an epoch. Nevertheless, for the sake of definiteness it is convenient to call that moment – at which the density of a *theoretical* universe will become infinite – the moment of the 'big bang' and measure the age of the universe from that moment. There are indications that the breakdown of Einstein's theory occurs at time-scales smaller than 10^{-44} seconds, when the temperature of the universe was higher than 10^{32} Kelvin. This is a *very very* high temperature and corresponds to an energy of 10^{19} GeV! At any reasonable temperature, like say 10^{10} Kelvin, or even 10^{20} Kelvin, Einstein's theory remains perfectly valid and can be used to make definite predictions. Using this theory, we can estimate the typical temperature of the universe at any given time. When the universe was about one second old, its temperature would have been about $10\,000$ million Kelvin. The average energy of a particle in the universe at this epoch is about 1 MeV, which is comparable to the energies involved in nuclear processes. The physical processes which operate at energy scales below about 100 GeV are well known from direct laboratory research. Hence we can certainly have faith in the predictions of the theory at these epochs. (We shall say something about the earlier epochs in section 5.4.)

What was the material content of the universe when it was, say, one second old? Note that the temperature of the universe at this time was so high that neither atoms nor nuclei could have existed at this time. Matter must have existed in the form of elementary particles. The exact composition of the universe at that time can be determined by tracking back the contents from the present day. It turns out that the universe at this time was a hot soup of protons, neutrons, electrons, positrons, photons and neutrinos. Among these particles, protons, neutrons and electrons are the basic constituents of all matter and are familiar to us in the present–day universe. Neutrinos are particles like electrons with no mass or electric charge, and interact only very weakly with the rest of matter (see chapter 2; section 2.3). For all practical purposes, we can ignore the existence of these neutrinos in our discussion. The positron (which is the antiparticle of the electron; see chapter 2; section 2.3) carries positive electric charge, while electrons carry negative electric charge. This unfamiliar particle existed in the early universe for a simple reason. The mass of the electron (or positron) is equivalent to an energy of about 0.5 MeV. When the universe was one second old, the temperature was about 1 MeV, which means that the energy of a typical photon was about 1 MeV. Since energy and mass are interconvertible, electron–positron pairs can be produced out of energetic photons at these high temperatures.

What about the number densities of each of these particles? This quantity can again be calculated from the present values of matter density and radiation density. To begin with, the total number of protons plus positrons should be equal to the total number of electrons. This is because the universe does not carry net electric charge. Each electron carries one unit of negative charge while each positron and each proton carries one unit of positive charge; only by keeping the total number of positrons plus protons equal to that of electrons will we be able to maintain overall charge neutrality. The number densities of protons and neutrons are approximately equal, with neutrons being slightly fewer in number. As regards the photons, note that both the number density of photons and that of protons decrease in the same manner as the universe expands. So the *ratio* of these two number densities does not change when the universe expands. If we know this ratio today, it would have been the same in the past. From the energy density of microwave radiation and the density of matter, we can compute this ratio today. It turns out that the number of photons in the universe is much larger than the number of protons or neutrons; to every nucleon there are approximately 10^9 photons in the universe. Finally, it is possible to compute the number density of neutrinos, given the number density of photons. It turns out that these are about equal.

Having decided on the contents of the universe when it is one second old, we can try to determine the evolutionary history of our universe. As the universe expands, it cools; the average energy of the photons will decrease and soon the photon energy will drop below 0.5 MeV. When this happens, the photons will become incapable of producing the electron–positron pairs. The existing electron–positron pairs, however, annihilate each other, producing radiation. When the universe cools to about 1000 million Kelvin, most of the positrons will have disappeared through their annihilation with electrons. We will now be left with a universe containing protons, neutrons, electrons and photons. Among these particles, protons are positively charged and electrons are negatively charged. They now exist in equal numbers, maintaining the overall charge neutrality of matter.

At this epoch, these particles exist as individual entities and not in the form of matter with which we are now familiar. To create the familiar form of matter, it is first necessary to bring protons and neutrons together and make the nuclei of different elements. Then we need to combine electrons with these nuclei and form neutral atoms. But we saw in chapter 2 that the typical binding energy of atomic systems is in the range of tens of electron volts. At the temperature of a few MeV such *neutral atoms* cannot exist. The binding energy of different atomic *nuclei*, however, is in the range of a few MeV. This suggests that it may be possible to bring neutrons and protons together and form the nuclei of different elements. However, this process – called 'nucleosynthesis' – is not easy. The key difficulty, of course, is the fact that the universe is expanding very rapidly at this epoch. (When the universe is one second old, it will double its size every few seconds; at present the universe is nearly 15 billion years old and hence it will double its size only in another 18 billion years. This shows why the expansion of the universe plays a major dynamical role in the early universe but not at present.) In order to bring particles together in one place we have to overcome the effects of expansion at least temporarily. This is quite impossible to achieve if we try to bring together several nucleons at once. For example, it is not easy to bring two protons and two neutrons together in one place to form the nucleus of helium, which is the second key element in the periodic table. Fortunately, there exists a nucleus called deuterium, made of one proton and one neutron with a binding energy of about 2.2 MeV. As the temperature falls below this value, we can combine protons and neutrons and form deuterium nuclei. And then, by combining two deuterium nuclei together, we can form the nucleus of helium. This suggests that the first heavy elements to form out of the primordial soup are deuterium and helium.

It is important to ask how much of these elements are produced. Note that the binding energy of helium is 28.3 MeV while that of deuterium is only 2.2 MeV. So it is energetically favourable for all the deuterium to convert itself into helium. This is roughly what happens; but the expansion of the universe prevents the process from actually reaching its logical end. A tiny fraction (about one part in 10 000) of the deuterium survives being converted into helium. It turns out that more of the deuterium survives if the density of protons plus neutrons (called collectively 'nucleons') in the universe is less. So, by measuring the amount of deuterium produced in the early universe, we can constrain the amount of nucleonic matter present in the universe (see figure 5.3). Deuterium is a fairly fragile nucleus and is easily destroyed during stellar evolution. So any deuterium we see in the universe must have been a relic of the early universe, though not all of it would have survived. Thus, observations give a lower bound to the amount of deuterium produced in the early universe (i.e., more could have been produced but not less). Observations suggest that this quantity is at least about 1 part in 10 000. Theoretical calculations, on the other hand, indicate that an abundance of 1 part in 10 000 is possible only if the total amount of nucleons in the universe is about 10^{-30} gm cm^{-3}. If the abundance of deuterium is more than one part in 10 000, then the amount of nucleons has to be still less. Thus you see that the *lower* limit on the observed

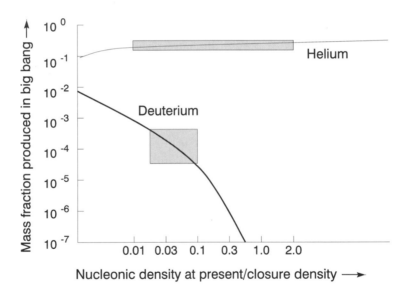

Figure 5.3 The amount of helium and deuterium synthesized in the early universe depends on the density of nucleons in the universe. The helium fraction is fairly insensitive to change in the density of nucleons, but the abundance of deuterium decreases rapidly with increasing nucleon density. The shaded boxes in the figure represent the observed values of helium and deuterium abundances. The observed deuterium abundance limits the density of nucleons to less than about one tenth of the critical density needed to close the universe (see section 5.3).

deuterium abundance leads to an *upper* limit on the amount of nucleons present in the universe. As we shall see, this result has important implications.

Let us consider next the amount of helium which would have been produced in the early universe. Since each helium nuclei has two protons and two neutrons, one requires equal amounts of protons and neutrons to form helium nuclei. If the universe has precisely equal amount of protons and neutrons, then all of them can combine to form helium. If for any reason there are a smaller number of neutrons, then these neutrons will combine with an equal number of protons, forming a certain fraction of helium nuclei, and the remaining protons will just stay as hydrogen nuclei. So the fraction of helium which is formed in the early universe depends on the ratio of number densities of neutrons and protons.

Originally, in the very early universe, the number densities of neutrons and protons are equal. However, as time goes on, the number density of neutrons decreases in comparison with that of protons. This is because free neutrons in the universe undergo a 'decay' called 'beta decay' in which the neutron converts itself into a proton and electron (actually it also produces an antineutrino in the process; but that is irrelevant for our considerations). Because of this beta decay, the number density of neutrons is continuously depleted vis-à-vis the number density of protons. By the time the universe has cooled enough to produce helium, there will be only about 12 neutrons to 88 protons in every collection of 100 nucleons (i.e., 12 per cent of the nucleonic mass is in neutrons while 88 per cent of the mass is in protons). These 12 neutrons will combine with 12 protons forming helium; the remaining 76 protons will end up as hydrogen nuclei. Thus, about 24 per cent of the mass will be in the form of helium nuclei, while 76 per cent of the mass will be in the form of hydrogen.

In principle, one could try to produce heavier elements by further nuclear reactions. In practice, however, it is not easy to synthesize these atomic nuclei in large quantities, mainly because the expansion of the universe prevents sufficient number of protons and neutrons from getting together. (Stars manage to synthesize heavier elements because they can do it over millions of years at sufficiently high temperatures. In a rapidly cooling universe this is not possible.) Detailed calculations show that only helium is produced in significant quantities. There will be very small traces of deuterium and other heavier elements.

The conclusion described above is a key prediction of the standard cosmological model. According to this model, only hydrogen and helium were produced in the early universe. All other heavier elements, which we are familiar with, must have been synthesized elsewhere. We now know that this synthesis took place in the cores of stars which had sufficiently high temperatures.

As time goes on, the universe continues to expand and cool. Curiously enough, nothing of great importance occurs until the universe is about 400 000 years old, by which time its temperature would be about 3000 K. At this temperature, matter makes a transition from the plasma state to the ordinary gaseous state, with the electrons and ions coming together to form normal atoms of hydrogen and helium. Once these atomic systems have formed, the photons stop interacting with matter and flow freely through space. When the universe expands by another factor of a thousand, the temperature of these photons will drop to about 3 Kelvin or so. *Hence we should see all around us today a radiation field with a temperature of around 3 Kelvin.*

We saw earlier that the universe indeed contains such a thermal radiation at a temperature of 2.7 K. In 1965, two scientists, Arno Penzias and Robert Wilson at Bell Labs, stumbled on the discovery of this radiation. A radio receiver they were developing detected microwave radiation which did not originate from any of the known sources in the sky. Further studies showed that this radiation was filling the entire universe and was reaching us — so to speak — from the depths of space. It was uniform all over the sky and had a characteristic temperature of about 2.7 Kelvin; it was as though the entire universe was enclosed in a box kept at this temperature. This thermal radiation covering the universe is a tell-tale relic of an earlier hotter phase. As we shall see later, observations related to this radiation play a vital role in discriminating between cosmological models.

5.3   Dark matter: weight without light

The description of the universe given in the last section contains the best understood features of conventional cosmology. You would have noticed that we started the discussion with a universe which was one second old and carried it forward to about 400 000 years. Though this is an impressive feat, it leaves three questions unanswered: What happened before one second? What happened to the universe between 400 000 years and now? What will happen to the universe in the future?

The first of these three questions is difficult to answer and has led to several theoretical speculations. We shall touch upon some of these speculations in sections 5.4 and 5.5. The second question could possibly be answered in a more convincing (and less speculative) manner, and will form the core of the sub-sequent chapters in this book. In this section we shall briefly discuss the third question; viz., what will happen to the universe in the future? Will it go on expanding for ever? Or will it reach a maximum size and start contracting afterwards?

This situation is analogous to the behaviour of a stone thrown vertically up from the Earth. Usually, the stone will rise up to a maximum height and fall back on Earth. If we increase the initial speed of the stone, it will reach a higher altitude before falling back. This, however, is true only if the initial speed is below a critical speed called the 'escape speed'. If the stone is thrown with a speed larger than the escape speed, it will escape from the gravity of the Earth and will never fall back. In other words, if the gravity is strong enough, it can reverse the speed of the stone and bring it back; but if the gravity is not strong enough, the stone will keep moving farther and farther away from Earth. We can compare the expanding universe to a stone which is thrown from the Earth. If the gravitational force of the matter in the universe is large enough, the expansion can be reversed and the universe will eventually start contracting. Since the gravitational force increases with the amount of matter, the fate of the universe depends on the amount of matter present in the universe today. If the matter density is higher than a critical value, this gravitational attraction will eventually win over the present expansion and the universe will start contract-ing. The critical value of density which is needed for this recontraction to take place can be estimated if we know the speed with which the universe is expanding – which in turn is related to the value of the Hubble constant. It turns out that for our universe to contract back eventually, it should have a matter density of about 5×10^{-30} gm cm^{-3} or more. This density is called the 'critical density' or 'closure density' of the universe, and will play an important role in our discussions.

So far, so good. Now we need to know the actual mass density of the universe. Is it higher than the critical density or lower? The future of the universe depends on the answer to this question.

A partial answer to this question is easy to obtain. We can make a fairly good estimate of the amount of matter which exists in the galaxies, clusters, etc., as long as the matter is emitting electromagnetic radiation of some form: visible light, X-rays or even radio waves. Telescopic observations covering the entire span of the electromagnetic spectrum, along with some theoretical modelling, allow us to determine the amount of *visible* matter which is present in our universe. The density contributed by this visible matter turns out to be less than about one tenth of the critical density needed to make the universe recontract. So if all matter in the universe is visible, then the universe will expand forever.

Unfortunately, this is not the whole story. It turns out that *not* all the matter which exists in the universe is visible! In fact, it could very well be that up to ninety-five per cent of the matter in the universe is 'dark' and does not emit any form of electromagnetic radiation. If this is the case, then it is important to identify and estimate the amount of dark matter which is present in the universe, since it is the dark matter which is governing the dynamics of our universe.

How do we determine the amount of dark matter present in the universe? Astronomers use different techniques for different objects, but the basic principle is to look for gravitational effects which are *unaccounted* for by the visible matter. Consider, for example, the spiral galaxies. (These are galaxies in which most of the visible matter is contained in a plane in the form of a circular disc; see section 4.6.) In the outskirts of such galaxies there exist tenuous hydrogen clouds orbiting the galaxy in the plane of the disc. If we see such a galaxy edge-on, then the hydrogen clouds at two sides of the galaxies will be moving in opposite directions – one towards us and one away from us. These clouds will be emitting radiation with a wavelength of 21 cm (see chapter 2; section 2.5) which – because of the Doppler effect – will be redshifted at one end and blueshifted at the other. By measuring these shifts, we can determine the speed of these clouds. Knowing the speed and the location of the clouds, one can estimate the amount of gravitating matter present in the galaxy influencing the motion of the cloud.

Astronomers were led to a very surprising conclusion when they performed the above analysis. It turns out that galaxies like ours contain ten times more mass than that determined from the stars and gas. That is, there exists 'dark matter' which is ten times more massive than the visible matter in the Milky Way! Studies carried out for different kinds of galaxies have repeatedly confirmed this result: all galaxies have large dark matter halos around them, with the visible part, made of stars, etc., forming only a tiny fraction. Most of the gravitational force in the galaxy is due to the dark matter, and hence the dynamics of the galaxy is mostly decided by the dark matter halo rather than by the visible matter.

There is another surprising thing about the dark matter: it seems to be distributed over sizes nearly three to four times bigger than the visible matter. In fact, astronomers have not been able to determine the 'edge' of the dark matter halo around any of the spiral galaxies. Remember that the inference about the dark matter was made based on the observations of the redshifted 21 cm radiation from the hydrogen clouds. These hydrogen clouds could be detected only up to some distance from the galaxy; so the observations regarding the dark matter halo cannot be pushed beyond the region where the hydrogen clouds exist. For some of the galaxies, this distance is as large as 70 kpc, while the size of the galaxy is only about 20 kpc or so. It is seen that there is a dark matter halo up to this distance of 70 kpc, and we have no information as to what happens beyond this distance. These numbers vary from galaxy to galaxy, but the overall impression one gets from all these observations is that the visible matter is only a tiny speck embedded in the middle of a vast dark matter structure.

What about structures larger than galaxies? How much mass, for example, is contained in the Local Group of galaxies? One can again add up the visible mass of all the galaxies to obtain the total visible mass. But each of the galaxies contains a dark matter halo around the visible mass. Since the physical size of these dark matter halos is quite uncertain, it is natural to assume that even groups of galaxies will contain significant quantities of dark matter in addition to the visible component. In the case of spiral galaxies, one could measure the amount of total mass by measuring the speeds of stars. It is not possible to use such a technique for systems like groups of galaxies which have irregular shapes. As a result, the mass estimate for groups is far more uncertain. One technique, which is usually adopted, is the following: from the Doppler shifts, one can measure the speeds of various galaxies in a group, thereby obtaining an estimate of the amount of kinetic energy per unit mass. It is possible to relate this quantity to the total mass and size of the system using some theoretical modelling. Since the size can be estimated fairly accurately, this method provides a handle for measuring the total mass of the group. Such measurements indicate that the dark matter in groups and clusters can be 10 to 100 times larger than the visible matter, even with conservative estimates.

On still larger scales, one can use the motion of clusters themselves to infer the distribution of gravitating matter. At the largest scales probed so far, the contribution from dark matter seems to be about half of what is required to make the universe recontract. But these observations also suggest that the dark matter density increases as we probe larger and larger scales. It is, therefore, quite possible that the total amount of dark matter contained in our universe is sufficient to reverse its expansion eventually.

What is this dark matter made of? At first one might have guessed that it could also be made of protons and neutrons like all the rest of matter. In principle, a galaxy could contain a large amount of mass in the form of Jupiter-like planets, say, which will be too faint to be detected by any of the telescopes. There is, however, an indirect piece of evidence suggesting that this is not the case. We saw in the last section that from observations of deuterium abundance it is possible to put strict bounds on the number density of nucleons which can exist in our universe. This bound implies that nucleons (which make up ordinary matter) could at best only account for ten per cent of the critical density. The amount of dark matter seen in various systems is far too large to be made of ordinary matter. Most cosmologists believe that the dark matter is made of more exotic particles – particles that interact only very weakly and have been produced in copious quantities in the very early phases of the universe.

For example, consider the elementary particle, the neutrino. We saw in chapter 2 that there are three kinds of neutrinos, corresponding to the three leptons, the electron, the muon, and the tau. These neutrinos are, of course, seen in the laboratory, and hence there is no doubt about their existence. What *is* doubtful, however, is the mass of these particles. All we know from laboratory experiments is that the masses of the electron neutrino, muon neutrino and tau neutrino are less than 12 eV, 250 keV and 35 MeV respectively; and of course, they could well be zero. The muon neutrino, for example, can have a tiny mass of about 30 eV without contradicting any of the laboratory observations. This mass is nearly 30 000 000 times smaller than the mass of a nucleon; but our universe contains nearly 300 000 000 times more neutrinos than protons. So the total mass density contributed by neutrinos can be ten times larger than that of nucleons, and can very well account for the dark matter.

Neutrinos are not the only possible candidate for the dark matter. Several particle physics models postulate the existence of more exotic particles called 'WIMPS' (Weakly Interacting Massive Particles), which could be far heavier than protons, but much less abundant. (If such a particle exists, we have to add that to the list of particles which populated the early universe, in our discussion in the last section.) There are currently several experiments in progress to detect such exotic particles, though none of them have succeeded yet. A laboratory detection of a dark matter candidate particle will revolutionize our understanding of the universe and establish a remarkable connection between the physics at the smallest and largest scales.

In summary, we may say that the universe contains 10 to 100 times as much dark matter, distributed at different scales, compared to visible matter. This dark matter is most likely to be made of exotic particles rather than ordinary protons and neutrons. Until one knows the exact quantity of dark matter present in the universe, it is not possible to predict whether the universe will expand forever or will reach a maximum size and collapse back.

There is another, related, aspect to the density of the universe. If the universe has a density equal to the critical density and is made of the normal kind of matter, then it turns out that it will have an age of about 13.6 billion years if $h = 0.5$ and only half that value if $h = 1$ (see page 99). The ages of the oldest stars in our galaxy seem to be about 12 billion years with an uncertainty of 2 billion years either way. Since we do not want the age of the universe to be less than that of the oldest stars, you can see that cosmological models with critical density can be in big trouble if h turns out to be closer to unity.

5.4   The very early universe

Let us now come to the more speculative questions among those raised at the beginning of the last section. While discussing the evolution of the universe, we started at a time of about one second. What happened before that? How did the universe get to this stage? Where did all this matter come from? These are the questions which had intrigued scientists and philosophers alike over the centuries, and at last we have some hopes of getting the answers.

The behaviour of the universe at earlier and earlier phases becomes more and more (theoretically) uncertain due to several reasons. As we go to very early phases of the universe the temperature is very high. This means that the elementary particles which constitute the matter in the universe are moving with very high energies. The interaction of particles and the laws governing their behaviour are uncertain when the energies involved are very high. Today, man-made accelerators in the laboratory can produce particles with energies of about 100 GeV. When particles have energies significantly higher than 100 GeV, we have no direct experimental evidence to understand their behaviour. Any theoretical model describing very high energy interactions is based on some broad symmetry principles rather than empirical facts. Given any such theoretical model, it is possible to work out the consequences for the evolution of the early universe. In other words, how the universe behaved during its infancy depends crucially on the particle physics models adopted for high energy interactions. Until the latter is settled, the former cannot be predicted uniquely.

The usual procedure is to construct different possible particle physics models for high energies and work our their consequences for cosmology, with the hope that one will be able to constrain the particle physics models from cosmological results. Thus, in principle, the early universe provides a 'poor man's particle accelerator', allowing one to probe high energy physics. In practice, however, this is not very easy because the predictions made by models may not be unique or may not be easily testable.

What are the features that make up a model describing particle interactions at high energies ? The key ingredient in these models is an attempt to describe all known interactions in a unified way, with a minimum number of parameters. At low energies, one notices that there are four different kinds of interactions between the particles: gravitational, electromagnetic, weak and strong. (See section 2.3.) The first two are familiar from everyday physics; the strong and weak interactions have very short ranges and operate in the interior of an atomic nucleus. They are responsible for several nuclear reactions as well as for phenomena like beta decay. As one probes higher energies, one realizes that these four interactions may not be independent of each other but could be connected together in a systematic manner. For example, at energies above 50 GeV, the electromagnetic and weak interactions merge together as a single interaction called 'electro-weak'. It is hoped that at still higher energies – which could be as high as 10^{14} GeV – the strong interaction can also be 'unified' with the electroweak force. At still higher energies, one expects gravity also to form part of the parcel, thereby leading to a single unified description of all the forces of nature.

Each of these forces is mediated by a particle; the electromagnetic interaction, for example, arises due to the exchange of photons, and the electroweak force arises from the exchange of particles called W and Z bosons (see chapter 2; figure 2.7). At high energies, these particles will also be produced in large quantities, and the dynamics of the early universe will be influenced by their existence. If the strong interactions are unified with the other forces at still higher energies, then it is necessary to have some more particles to mediate these unified interactions. These particles, (called X and Y bosons) are predicted to exist (at high energies) in some of these theoretical models. While the W and Z bosons are seen in the laboratory, we have no direct evidence for the X and Y particles. However, their properties can be predicted if the theoretical model is known.

There are some interesting new features which are brought into cosmology by such particle physics models. Consider, for example, the asymmetry which exists between matter and antimatter. Most of the universe is made of protons, neutrons and electrons, but we do not see a corresponding number of anti-protons, antineutrons and positrons. This is surprising because at low energies, the interactions are symmetric between matter and antimatter. For example, the strong nuclear force which acts between a proton and another proton is the same as that which acts between a proton and an antiproton or between an antiproton and an antiproton. That is, nuclear forces seen in the laboratory do not care whether the particle is a proton or an antiproton. If we start with equal amounts of matter and antimatter in some region of the universe, then these interactions cannot make them unequal, since these interactions do not distin-guish between matter and antimatter. So how is it that our universe has matter but not antimatter? We can say 'That is the way the universe is!' but that is no explanation at all! It would be nicer if we could start off the universe with equal amounts of matter and antimatter and induce an asymmetry through some dynamical process. Since the asymmetry already existed when the universe was, say, one second old, something which happened earlier on should have triggered this asymmetry.

It turns out that some of the models which unify the strong interactions with electro-weak interactions do require forces which break the symmetry between matter and antimatter. If these models are correct, then it is possible to start with a perfectly symmetric universe and arrive at a situation in which there is more matter than antimatter. Of course, these models will also allow us to compute the amount of excess matter over antimatter, and this could serve as a crucial test for such particle physics models. Unfortunately, the models are still not sufficiently well-founded to predict a single unique number for this quantity. By varying the parameters in these models, it is possible to get any reasonable value. Nevertheless, it must be remembered that the influx of par-ticle physics ideas into cosmology has now at least provided us with a proce-dure by which such questions can be asked and answered.

The models for high energy particle physics also predict different scenarios for the evolution of the earliest moments of the universe. Some of these scenarios contain pleasant surprises which help one to understand certain puzzling features of our universe. One such scenario which shot into promi-nence in the last decade is called the 'inflationary universe". This scenario has helped cosmologists to look at the universe in an entirely new light.

To understand the idea behind the inflationary universe and its importance, it is first necessary to recall some simple features of the expanding universe. If the universe is dominated by radiation, then it expands by a factor of 2 when the time elapses by a factor of 4, expands by a factor of 3 when the time elapses by a factor of 9, etc. Such an expansion is called a 'power law' expansion and leads to certain conceptual difficulties.

The first difficulty is the following: in an expanding universe, physical processes cannot act as effectively as in a static universe. The following analogy will be helpful in understanding this feature: suppose we divide a container into two parts and keep gases at different temperatures on the left and right halves. We now remove the partition and let the gases mix. Within a very short time the hot and cold gas will mix with each other and a common temperature will be reached inside the container. But suppose the walls of the container were expanding when the gases were trying to mix. Then the equality of temperature cannot be reached so easily. Regions in the centre of the container, where hot and cold gases are close to each other, will attain equal temperature after some time. But whether the entire amount of gas will attain a constant temperature in an expanding container depends on two factors: (i) the rate at which gas molecules can interact with each other and exchange energy; and (ii) the rate at which the container (and the gas) is expanding. If the rate of expansion is much higher than the rate of interaction, temperature equality will not be reached and the final state of the gas will be made of different patches with different local temperatures. It turns out that in the radiation dominated universe the expansion rate is such that no significant mixing can take place. In other words, if regions of the universe have different temperatures to start with, then they will never come to equilibrium and reach the same temperature. What is more, the patchiness of the temperature of the radiation must be visible in the sky, in the microwave radiation. Observations, however, show that the temperature of the microwave radiation is almost the same in all directions, suggesting that the temperature in the early universe was the same at all spatial locations. In the conventional cosmological models this is a bit of a mystery.

There are two ways out of this difficulty. The first one, of course, is to slow down the expansion tremendously – in the early phases of the universe – so that sufficient mixing can occur. This, however, is not possible because we know the rate at which the universe is expanding today and, from this value, we can calculate how fast it would have been expanding in the past. It is not slow enough for this idea to work. The second way out is to let the universe expand very fast for a short time, after some amount of mixing has take place. In this process, we take a small region of the universe which has some constant temperature and expand it by a very large factor to form the currently observed region of the universe. The universe *is* very patchy in such a model, but the region which we observe today is made out of a uniform patch, expanded by a large factor. Such models of the universe – which make the universe expand much more rapidly than the conventional models – are called 'inflationary models' and the rapid phase of expansion is called 'inflation'.

What drives the universe into such a phase of rapid expansion? Normally it is the energy density of the radiation that causes the early universe to expand. As the universe expands, this energy density decreases and consequently the expansion rate also slows down. During the inflationary phase, however, the energy density remains constant (thereby maintaining a constant rate of expansion) even though the volume is increasing. In order to achieve this, the inflationary phase of the universe should contain matter with negative pressure. In normal systems which we are accustomed to (like a balloon filled with gas), pressure is a positive quantity. If the balloon expands, the energy content and pressure inside the balloon will decrease. But for a system with negative pressure the energy content will stay constant (or even increase) if the system expands! So, for inflationary models to work, we need some kind of matter with negative pressure. Normal matter, of course, does not have this property. But quantum theory comes to one's rescue in this case. We saw in chapter 2 that quantum theory establishes a correspondence between particles and wave fields. For example, the particle photon has an electromagnetic field associated with it. In the same way, some of the esoteric particles, needed to unify strong and electro-weak interactions, have some purely quantum-mechanical fields associated with them. It turns out that certain models of particle physics use quantum fields which can have negative pressure (in the sense explained above). What inflationary models do is to arrange matters in such a way that these quantum fields dominate the expansion of the universe for a short span of time in the very early universe.

There is another significant fall-out from the inflationary models for the universe. The present day universe contains several inhomogeneous structures like galaxies, clusters, etc. It is believed that these structures have originated from the growth of small inhomogeneities present in the early universe. (We shall say more about this in the next chapter.) In order to develop a complete theory of structure formation, one has to also understand how these small inhomogeneities came into being in the first place. This problem is not adequately answered in conventional cosmology. In fact the expansion rate of the radiation dominated universe is not conducive to producing inhomogeneities of the correct type. Inflation of the universe helps in this regard as well. The quantum fields which were driving the inflation will also contain small fluctuations in energy density in a natural manner. These small fluctuations can grow and form the seeds for structures like galaxies. Given any particle physics model which can produce inflation, one can also compute the kind of fluctuations that would have been generated. This feature can help in understanding the formation of structures in the universe.

How can we test such models? Rather, how can we be sure that the universe ever underwent an inflationary phase ? Since the inflation is supposed to have occured at a very early phase of the universe (at about 10^{-35} seconds after the big bang!), it is not possible to test these models by direct observations. The typical energy of particles in the universe during inflation is far higher than the energies achieved in the laboratory; hence it is also not possible to verify the particle physics models leading to inflation directly in the laboratory. However, some indirect verification is possible. It turns that generic inflationary models predict that the universe will have a density which is equal to the critical density. So if the observations conclusively show that the density of the universe is less than the critical density, then we can rule out the possibility that our universe underwent an inflationary phase. Unfortunately, we do not have sufficiently accurate measurements of the total matter density of the universe to use this as an effective test. Each inflationary model also predicts a particular pattern of fluctuations for the inhomogeneities of the universe. The microwave background radiation contains information about these fluctuations. By comparing observations of the microwave radiation with predictions of the inflationary models, one can validate or rule out these models. Unfortunately, the observational data on the microwave background radiation is not of sufficiently high quality to pin down the models of inflation. This situation will probably improve in a few years. Most of the models of inflation also predict the existence of a uniform background of gravitational wave radiation in our universe. The intensity of these gravitational waves, however, is too low to be detected by existing technology. But in principle, with an improved technology for gravitational wave detection, one should be able to test this prediction from inflation. observational tests of

Thus the observational situation regarding inflation is quite unsatisfactory at present – we have no evidence that the universe underwent an inflationary phase. Nevertheless, many theoreticians believe that inflation is a good thing (for once!) and is here to stay.

5.5 The quantum universe: science of genesis

In studying the very early universe, we have introduced the concept of a 'big bang'. The big bang was the moment at which the size of the universe was zero and the density of matter was infinite. It almost appears as though the universe was 'created' at this instant. What does it mean? How does one understand scientifically the epoch of the big bang?

An instant at which the universe is infinitely hot and infinitely dense seems an ideal candidate for the moment of 'creation'. In fact you often come across sentences in popular (and, alas, even in technical) literature running like this: 'the universe was created 15 billion years ago in a big bang'. What is conveniently forgotten is that the universe was also *infinitesimally* small at this time – very much smaller than the size of an atom! In trying to understand such a phenomenon, we are extrapolating Einstein's theory of gravity, and the cosmological models based on it, to very *small* distances. However, the physics of the microworld needs to be described in the language of quantum mechanics. Until we produce a quantum theoretical description of cosmology, we cannot understand the very early moments, close to the big bang. A branch of science called 'quantum cosmology' attempts to provide such a description and could possibly have some answers.

The quantum mechanical description of nature is very different from the classical description of everyday physics. Consider, for example, the concept of a trajectory. A pool ball hit by the cue will follow a well defined trajectory. The laws of physics can be used to predict the exact position and speed of the pool ball at any later instant. In the microworld of electrons and atoms no such description is possible. One of the key concepts in quantum theory, which describes the physics of the microworld, is called the 'uncertainty principle'. According to the uncertainty principle, which forms the backbone of quantum theory, the position and speed of a particle cannot be determined simultaneously and exactly. If we know the position of an electron exactly at one instant, then we will not know its speed at all; hence we will not know where the electron will be in the next instant of time. So the concept of a trajectory fades into insignificance. Quantum theory replaces it with a new concept: that of the 'quantum state'. Instead of saying an electron follows a particular path, we merely say that the electron is in a particular quantum state. In general, a particle will not have either a definite speed or a definite position in such a state; instead, we can only compute the *chance* that the particle may be found in one location or another; or moving with one speed or another.

This transition from a 'deterministic description' (in which the behaviour of everything is precisely determined and completely predictable in principle) to a 'probabilistic description' (in which only the chances or probabilities can be predicted) is the major new feature brought in by quantum theory. Such a description, based on probabilities, introduces the concept of 'quantum fluctuations'. According to classical physics, a pool ball can possess exact values for its energy, speed, position and so on, all simultaneously. According to the quantum mechanical viewpoint, such certainty is not possible. A particle with a definite energy will not (in general) have a unique position; each attempt to measure its position will give a different result. There will be an average position, but there will also exist inevitable 'quantum fluctuations' about the average position.

To understand the physics near the 'big bang', we need to understand how the quantum mechanical concepts modify our description of the universe. The main feature of classical cosmology is the expansion of the universe, which is described by a particular function of time called the 'expansion factor'. Suppose we are located in a particular galaxy and we study the position of any other galaxy. As time goes on, this position changes due to the expansion of the universe, and the specific manner by which it changes is decided by the expansion factor. (This is quite similar to the manner in which the position of a billiard ball changes during its motion, which is what we call the trajectory of the ball). It starts at zero at zero time and increases with time. The exact form of the expansion factor depends on the nature of the matter populating the universe.

Classically, the expansion factor is governed by a set of equations first formulated by Albert Einstein. These equations allow one to determine the expansion factor *and* the rate at which it is changing at any given moment. Given these two pieces of information, one can determine the value of the expansion factor at the next instant. This situation is like knowing both the position and the speed of a particle at a given instant, so that one can fix its trajectory. To produce a quantum theory of the universe, one can begin by providing a quantum description for this expansion factor. Quantum theory will, of course, prevent us from knowing simultaneously both the expansion factor and its rate of change (just as it prevents us from knowing both the position and speed of a particle) simultaneously. This uncertainty, of course, it quite irrelevant when the universe is bigger than atomic dimensions, but it does change things drastically when we deal with the very early moments close to the big bang.

In a quantum-mechanical description for the universe, we have to replace the concept of an expansion factor (which is like the concept of a trajectory for an electron) by the concept of a quantum state for the universe (which is like the state in which the electron has a definite energy in the hydrogen atom). As we said before, the position of the electron will not have a definite value in a state in which the electron has a given energy. All we can talk about is the chance for the position of electron to have any particular value. There will be an average position, but there will also exist fluctuations around this average position. Similarly, when the universe is described by a particular quantum state, its expansion factor will not have a definite value. One can talk about the average value of the expansion factor, but there will be fluctuations around this average value. These fluctuations will become larger as we go to smaller and smaller values of the expansion factor. For this reason, the quantum-mechanical description of the universe has a hope of avoiding the 'big bang' at which the size of the universe becomes zero.

The situation here is, in fact, quite similar to the physics of a hydrogen atom. Since the proton and electron are oppositely charged, there is a force of attraction between the two and – classically – the electron will move under this attractive force and fall on the proton in a very short time. This fate of the electron was known before the days of quantum theory, and people considered the stability of hydrogen atom to be a deep mystery. Quantum mechanically, however, a stable hydrogen atom consisting of a proton and an electron can exist. This is because, in quantum theory, the uncertainty principle prevents the electron from coming too close to the proton; for if it were to come very close there would be insufficient uncertainty about its position and speed. Another way of stating the same result is the following: classically, the electron moves in a definite trajectory which has zero size after some time; that is, the electron falls to the centre where the proton is located. Quantum mechanically, the electron does not have a definite trajectory, but instead could exist in some energy level. In a given energy level, the electron does not have a definite position, and we can no longer talk in terms of trajectories which make the electron fall on the proton.

Similar ideas work in quantum cosmology. The dynamics of the expansion factor can be expressed in quantum mechanical language, leading to a description that is similar to a quantized hydrogen atom. Just as the electron is prevented from approaching too close to the proton (leading to zero dimension for the hydrogen atom), the universe is prevented from reaching zero size due to quantum effects. It turns out that the smallest size the universe can attain, according to this theory, is of the order of 10^{-35} meters. This size, called the 'Planck length', can be made entirely in terms of certain fundamental constants in physics.

Given a description of a quantized universe, we should be able to ask questions about the 'creation'. The most important question is: 'Why is there a universe at all? Why is there all this matter?' Classical cosmology offers us no explanation: given a certain amount of matter in the universe, classical cosmology merely traces it back to the big bang and then throws up its hands, being at a loss to explain further! An empty and uneventful space is as acceptable to classical theory as all the richness of our evolving universe.

Some of the quantum mechanical models of the universe do give an answer. One can show – in these models – that equations describing such a universe *do not allow* an empty universe. In other words, quantum cosmology not only describes a universe with matter in it, but actually demands that there should be matter in it. In these models, empty space becomes unstable due to the quantum fluctuations of the gravitational field and makes a spontaneous transition into an expanding universe filled with matter. You may think that creating matter and making it expand would require some amount of energy, and hence if matter came out of 'empty space' we would be violating the conservation of energy – which is a cherished principle in physics. In fact, this is *the* question which had plagued several philosophers: how can something emerge from nothing? Actually there is no violation of energy conservation in this process. Remember that the gravitational field has negative energy because it can bind matter together. (We saw earlier that the electron bound to a proton exists in a state with negative energy; any attractive force will share this feature and gravitational force is always attractive.) So we can easily come up with a situation in which the negative gravitational energy exactly balances the positive energy of matter (and expansion), thereby keeping the total energy zero. The equations of quantum cosmology indicate that the universe has essentially sprung spontaneously from the vacuum due to quantum fluctuations.

This might seem a very bizarre phenomenon unless you properly grasp the subtleties of quantum-mechanical descriptions of the universe. The vacuum of quantum theory is not the 'mere emptiness' of classical theory. Instead it is bristling with elementary particles constantly being created and destroyed, and fields which fluctuate incessantly. The mathematics of these models indicate that such a state of affairs *can* lead to the creation of a material world.

It would, of course, be very wrong to say that all that needs to be known is known. At best, quantum cosmology is in its infancy. Several links in the chain of arguments need to be strengthened and several gaps need to be filled. All the same, combining quantum theory and general relativity – two major intellectual triumphs of this century – seems to lead us to a far richer and more beautiful vision of our universe.

6

Forming the galaxies

6.1 Gravitational instability

We saw in chapter 4 that our universe contains a hierarchy of structures from planetary systems to superclusters of galaxies. Between these two extremes we have stars, galaxies and groups and clusters of galaxies. Any enquiring mind will be faced with the question: how did these structures come into being?

Is it possible that structures like our galaxy have always existed? The answer is 'no' for several reasons. To begin with, stars are shining due to nuclear power which runs out after some time. So it is clearly impossible for any single star to have existed for infinite amount of time. One can, of course, recycle the material for a few generations but eventually even this process will come to an end when all the light elements have been exhausted. So clearly, no galaxy can last for ever. Secondly, we saw in chapter 5 that – at the largest scales – the universe is expanding, with the distance between any two galaxies continuously increasing. This led us to a picture of the early universe with matter existing in a form very different from that which we see today. It follows that the structures like galaxies which we see today could not have existed in the early universe, which was much hotter and denser. They must have formed at some finite time in the past.

How early in the evolution of the universe could these structures have formed? As we shall see, we do not have a definite answer to this question. However, we saw in the last chapter that neutral gaseous systems formed when the universe was about 1000 times smaller. It is reasonable to believe that the structures like galaxies must have formed between the redshift $z = 1000$ and the present moment. The key question, of course, is *how* did these structures form? After all, one can think of a universe containing a completely homogeneous mass of gaseous hydrogen and helium, expanding steadily for ever. In order to form the structures, it is important for the gas to become more concentrated and denser in some regions compared to some other regions.

For example, the mean density of matter inside a galaxy is about 10^{-24} gm cm^{-3}, which is nearly a million times larger than the average density in the universe. So some of the gas must have clustered together, increasing the density enormously over the background density, in order to form systems like galaxies. We need a mechanism which will convert a completely uniform distribution of gas into a distribution punctuated by clumps of matter.

This mechanism happens to be just our old friend, gravitational force. The gravitational force can help the growth of non-uniformity in any material medium, as illustrated in figure 6.1. The top part of this figure shows the density of a nearly uniform medium. The dashed line represents the average density, and the wiggles around the dotted line represent the actual density which is slightly higher in some places (like A) and slightly lower than average in locations like B and C. The wiggles in the density are expected to be very small compared to the average density. (They are exaggerated in the figure for clarity.) Consider now the gravitational force near A and B. Since the region around A has slightly more mass than the average, it will exert more than average force on the matter around it. This will make the matter around A contract, thereby increasing the concentration of matter around A. In other words, density at regions like B and C will decrease further while the density at A will go up. This is shown in the lower frame which depicts the distribution of density at a later time. Now there is even more matter around A; so the gravitational force near A will be still larger, causing still more matter to fall on to it, etc. Clearly, this process will make overdense regions grow more and more dense just because they exert greater influence on the surroundings. This process goes under the name 'gravitational instability', and is similar to the social situation in which the rich get richer and the poor get poorer.

In describing the process of gravitational instability, we have oversimplified matters a little bit. It might appear that the lump at A grows essentially by accreting matter from the surroundings. Actually, there is another – more important – process; since the region around A has greater than average density, it will also contract under its own weight. That is, the self-gravity of the lump at A will make it collapse further and further, which will also increase the density of the lump. (You see that the rich get richer by their own efforts as well as by robbing the poor.) In fact, this process is really the dominant cause for the increase of density at Λ.

While gravitational instability is the key mechanism for structure formation, it does not apply directly at all levels. Or, rather, other physical processes can play a more important role at some scales. By and large, cosmological models

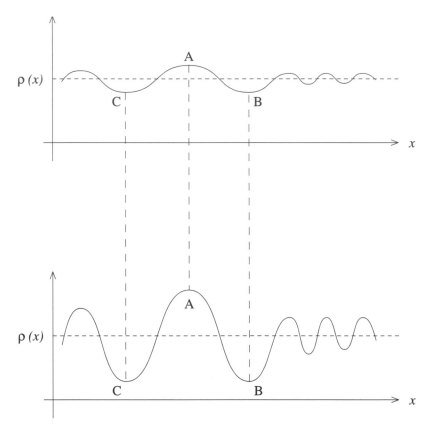

Figure 6.1 The process of gravitational instability is schematically depicted in this figure. The top frame shows the variation of density of a gas in space. The dashed line represents the average density and the wiggles around the dotted line represent the actual density, which is slightly higher in some places (like A) and slightly lower than average in locations like B and C. A high density region, like A, will contract under its own self-gravity and will also accrete matter from nearby regions like B and C. This process will make the lump at A denser and will decrease the density around it. So, at a later time, the density distribution will be as shown in the lower frame. This process, which leads to the growth of density inhomogeneities, is called 'gravitational instability'.

try to use gravitational instability to produce structures which are progenitors of galaxies i.e., lumps of gas with masses of about $10^{11}M_{\odot}$ distributed over a size of about 20 kpc. (The symbol $M_{\odot}$ stands for one solar mass.) Larger structures like groups and clusters of galaxies can form through the mutual gravitational attraction of the galaxies. Similarly, lumps of gas contained within the galaxy can contract under their own weight, eventually leading to the formation of stars. Stellar evolution synthesizes heavier elements, and supernova explosions distribute these elements within the galaxy. Density inhomogeneities in this material can lead to the second (and later) generation of stars like our Sun, as well to planets around them.

This picture suggests that the key unit which one should try to form is the galaxy, and smaller and larger structures could arise through the processes described above. Hence, most models for structure formation concentrate on forming galaxy-like objects. If we start with a gaseous system with small inhomogeneities in density, then gravitational instability can make the inhomogeneities grow in size, eventually leading to the formation of structures like galaxies, etc. But before we conclude that this is a viable process and that the structures in the universe *did* form by this process, we have to circumvent several vital difficulties.

To begin with, we have to ensure that there are no other effects counteracting *against* gravitational instability. When the gas cloud is compressed by its own gravity, its volume will decrease and – consequently – the gas pressure will increase. If the pressure can rise to a value high enough to balance the gravitational force, then further contraction will be halted and the process of gravitational instability will come to a grinding halt. Secondly, it must be noted that the expansion of the universe has the effect of pulling particles apart from each other; so, it acts against the direction of gravitational contraction. If the expansion is very rapid, then gravitational instability cannot be effective in increasing the density by a large factor. Thirdly, the picture of gravitational instability depended on the existence of small wiggles in the density at some time in the past. If you start with a strictly uniform density – with no lumps like A in figure 6.1 – then, of course, it will remain strictly uniform in the future. So we need to have a mechanism which will produce wiggles in the density distribution in the early universe. In technical language, one says that we must have a mechanism for producing 'density fluctuations'. These wiggles could, of course, be quite tiny; gravity can make them grow eventually but gravity cannot operate if there are *no* wiggles to begin with. Finally, it is important to have density fluctuations of the right kind. We shall discuss this problem in much greater detail later on, but the key idea is the following: the structures like galaxies, etc. which we see today have come out of the initial wiggles in the density. So clearly, if we change the initial pattern of wiggles we will end up with a different kind of galaxy distribution in the present universe.

In order to reproduce correctly the universe which we live in, it is important to have initial fluctuations of the correct kind.

Most of the current research in cosmology revolves around the questions raised in the last paragraph. It is generally accepted that gravitational instability is the primary cause of structure formation. But to make this mechanism work we have to get several processes to cooperate with each other in a specified manner. This task is far from easy and researchers are still racking their brains over it.

6.2   Testing the paradigm

The description in the last section attempts to build the entire universe starting from tiny density fluctuations which existed in the earlier phase. You could very well ask whether this model can be verified observationally or whether it is only a theoretical speculation. Fortunately, the gravitational instability model lends itself to a rather stringent test, which it passes – well, almost!

The key to testing the gravitational instability model lies in the microwave background radiation mentioned in section 5.2. We saw that when the universe was less than a thousand times smaller than at present, matter existed in the form of a plasma containing positively charged nuclei of hydrogen and helium and negatively charged electrons. The photons in the universe were strongly coupled to matter and were repeatedly scattered by the charged particles. When the universe cooled further, the electrons and nuclei combined to form neutral atoms and the photons decoupled from the matter. (This epoch is called the 'epoch of decoupling'.) These are the photons which we see around us in the form of microwave radiation. According to this picture, this radiation is coming to us straight from the epoch of decoupling. It follows that this radiation will contain an imprint of the conditions of the universe at that epoch; by probing this radiation closely we will be able to uncover the past.

If structures which we see today formed out of gravitational instability, then small fluctuations in the density of matter must have existed in the early universe and – in particular – around the time when the universe was a thousand times smaller. Since the radiation was strongly coupled to matter around this time, it would have inherited some of the fluctuations in the matter. What does one mean by fluctuations in radiation? Since the thermal radiation is described essentially by one parameter, namely, the temperature, fluctuations in the density would correspond to fluctuations in this temperature. Thus we reach two important conclusions: Firstly, if the universe was *very* inhomogeneous at the epoch of decoupling, then the microwave radiation field which we see today will also be *very* inhomogeneous. Secondly, if there were *small* fluctuations in the density field at the epoch of decoupling, then the microwave radiation should also exhibit similar fluctuations in its temperature.

We saw in section 5.2 that the microwave background radiation was found to be extremely isotropic; that is, its temperature was the same in all directions in the sky. Suppose, for a moment, that the picture for structure formation described in the last section was all wrong and that structures like galaxies had always existed. Then the microwave radiation should show significant fluctuations in the temperature across the sky. Since this is not seen, we may conclude that the universe was fairly homogeneous at the epoch of decoupling. It is this uniformity in the density of the universe, at the epoch of decoupling, which translates into the isotropy of the microwave radiation across the sky. (By 'isotropy' we mean that the radiation has the same properties in all directions.)

That, of course, is only a part of the story. For the gravitational instability to operate we *do* need some tiny fluctuations in the matter density at the epoch of decoupling. These 'seeds' — so to speak — will manifest themselves as small deviations in the temperature of the microwave radiation. Unfortunately, these deviations are too faint to be seen easily. Calculations showed that this temperature difference is extremely tiny — about one part in hundred thousand! But it is vital to look for these temperature fluctuations to verify the theoretical models.

Over the past two decades several experiments have attempted to measure these temperature variations. These experiments are difficult to carry out from the ground because of the interference due to water vapour in the atmosphere. Water vapour absorbs and emits radiation in the microwave region of the spectrum in which one is trying to make measurements. To have a fighting chance, one has to carry out experiments at exceptionally high or dry locations. Even then, the earlier versions of the experiments could only reach a sensitivity of three parts in hundred thousand and did not see any deviation. The microwave radiation looked quite uniform.

The real breakthrough came when NASA put into orbit a satellite, called the Cosmic Background Explorer (COBE), specifically designed to look for temperature deviations in the microwave radiation. The satellite experiment has a clear advantage over ground based observations in sensitivity. The hope was finally realized in 1992. Analysis of data collected by COBE has shown that the radiation field does have temperature variations of the expected magnitude: about one part in a hundred thousand!

The discovery has caused considerable excitement among cosmologists and – not surprisingly – in the popular press. It shows conclusively that the theoretical models about the formation of the structures in the universe are basically sound. The universe was indeed smoother in the past but not completely so. The small departures from uniformity, seen by COBE, could have grown into the galaxies we see today. This is a direct test of the gravitational instability picture and is in broad agreement with the ideas described in section 6.1.

In fact, the COBE measurement has given us a lot more. It tells us the magnitude of the density fluctuation as well as its broad pattern. We shall see in later sections how both of these can put restrictions on cosmological models.

6.3   Density contrast: the ups and downs

Let us now probe in detail how the mechanism of gravitational instability operates. To do this, it is convenient to introduce a quantity called 'density contrast'. Imagine a cube of size R kept at some region of the universe. The volume of such a cube is R^3, and if the average background density of the universe is ρ_{bg}, then this cube will contain, on the average, a mass $M_{bg} = \rho_{bg}R^3$. If the universe was completely uniform, then the density of the universe would have been ρ_{bg} everywhere and our imaginary cube will contain the mass M_{bg} irrespective of where it has been placed. But the real universe, as we have seen, contains fluctuations in density. Some regions will have density larger than the background density ρ_{bg} and some regions will have density less than the background density ρ_{bg}. Suppose we happened to keep our cube on a clump of size bigger than R and density higher than average. In that case the cube will contain a mass M which will be larger than M_{bg}. If we move the cube to another location, we might end up keeping it in a region which has density smaller than average, and then it will contain a mass less than M_{bg}. So, as we keep moving our cube around, the mass contained within the cube will fluctuate up and down around the value M_{bg}. If the fluctuation in the mass contained in a cube of size R is large compared to the mean value of mass M_{bg}, it is an indication of a large amount of inhomogeneities being present in the universe at a length scale of R. We define a quantity, called the 'density contrast', as the ratio between the fluctuations in the mass to the background mass. For example, a density contrast of 0.2 at a scale R would mean that the fluctuations are about 20 per cent of the background mass contained in the scale R. Suppose we are interested in the scale of galaxies which contain a mass of about $10^{11} M_{\odot}$. The density contrast of 0.2 at this scale would mean that the fluctuation in the mass is 20 per cent of $10^{11} M_{\odot}$; i.e., the fluctuations would be about $2 \times 10^{10} M_{\odot}$. By the same token, a density contrast of, say, 100 would mean that the scale under discussion has a very large fluctuation in mass.

The concept of density contrast is vital to everything which follows in this chapter. Hence, it may be worthwhile to illustrate it in terms of a more familiar example. Consider the price of a particular share in the stock market which you bought at a face value of £10. You will find that the price of the share varies on a daily basis. Assuming that the market is reasonably stable, you may find that the price of the share (with a face value of £10) varies as 10.01, 9.98, 10.03, 10.05, 9.95, 9.99, 9.99... on a set of consecutive days. If you calculate the average of these numbers you find that it is actually 10. So you might conclude that on the average the price of the share is neither going up nor coming down. But obviously the price is not exactly equal to £10 on all days but is fluctuating around it. Consider now another set of share prices given by 11.5, 10, 11, 8.5, 12, 10, 7 on consecutive days. Once again, if you calculate the average of these numbers you find that it is 10; so the price of the share is again not going up nor coming down on the average. But a comparison of the two sets of figures clearly shows that in the second case the prices are fluctuating a lot more than in the first case. One way of seeing this is to look at the difference between maximum and minimum share prices in the two cases. In the first set it was $10.05 - 9.95 = 0.1$, while in the second case it is $12 - 7 = 5$. Since the average price of the share was £10, we may say that the 'price contrast' is given by the fractional fluctuation in the price, which, in the first case was $(0.1/10) = 0.01$ or 1 per cent. In the second case it is $(5/10) = 0.5$ or 50 per cent! This situation is completely analogous to the density fluctuation in the universe. There are regions of higher density and regions of lower density (just as there are days when the share prices are higher than the average and days when share prices are lower than the average) with the average density being equal to that of our background universe. The fluctuation around the average can be described by a density contrast (which is analogous to the 'price contrast' defined above); larger density contrast indicates larger fluctuations in the density.

The share market example can also be used to illustrate some other features. Consider a period when the market is plagued by serious uncertainties. In that case it is possible for the share prices to fluctuate wildly, with the fluctuations increasing over a period of time. In other words, even though the average price of the share remains £10 the price contrast can grow with time. To see this, you only have to imagine a two week period with the prices given by 10.01, 9.98, 10.03, 10.05, 9.95, 9.99, 9.99, 11.5, 10, 11, 8.5, 12, 10, 7. (This set is obtained by combining the two cases discussed above.) The average price of the share remains £10 during the entire fortnight. But, clearly, the prices have fluctuated much more wildly during the second week than in the first week (making you, the share holder, a bit more nervous). Mathematically speaking, the price contrast has gone up from 0.01 to 0.5 in a week's time. The same phenomenon takes place during the growth of gravitational instability. The average density of the universe is not affected by the formation of structures. This is because, during the formation of galaxies, etc. matter is neither created nor destroyed; it just redistributes itself. (Of course, since the universe is expanding the average density is continuously decreasing, but that has nothing to do with the formation of structures.) The density contrast, however, keeps increasing, just as in our example the price contrast increased from the first week to the second.

In general, even the average price of a share will change with time. (After all, you invested in it hoping that the average price would go up.) When both the average price and the fluctuations change in time, the price contrast could very much depend on the timescale over which it is computed. In the above example, suppose we decide to do the calculation in every five day period and assume that the price was £10 on the fifteenth day. During the first five days the average price was £10.004 and the fluctuation was 0.01; during the second lap of five days, the corresponding figures were 10.5 and 0.15, and during the last five days we have 9.5 and 0.5. Calculating the price contrast as before, we get the numbers 0.01, 0.15 and 0.5. But when we computed the price contrast over the first week and second week we got the numbers 0.01 and 0.5. This leads to an important conclusion. The fluctuations depend very much on the scale over which we are studying the phenomenon. There are, of course, several other familiar examples which illustrates the same fact. If you look at a black and white photograph in a newspaper it will appear reasonably smooth. But if you look at the same photo through a magnifying glass you can see a set of black and white dots with much larger fluctuations in intensity. The coastline of a country is another example. Viewed from a satellite, it appears fairly smooth, but when you go to a nearby beach you can see much larger variations around the mean boundary. All these go to show that physical parameters can fluctuate rapidly at small scales, but when viewed over a much larger scale they can appear smoother.

Returning to the density fluctuations in the universe, the situation is again the same. Here the scale we are probing is decided by the size of the cube which we use. If we use a small cube, we may find that there are violent fluctuations in the mass as we move the cube around. But if we use bigger and bigger cubes, then the density contrast will become smaller and smaller. In other words the universe appears smoother and smoother when viewed at larger and larger scales. Thus we can characterize the inhomogeneities in the universe nicely by giving the density contrast at different scales. This quantity is of central importance in models for structure formation, and both theoreticians and observers struggle hard to determine it.

The discussion above centered around the fluctuations in density at any given moment in time. As the universe evolves, we expect the fluctuations to grow due to gravitational instability. Just as the pattern of fluctuations at any given time could be different in different models, it will also look different at different times. To understand the complexities of structure formation, we therefore have to understand the various factors in which the density fluctuation can depend on and the rate at which it grows. Let us now take a look at these factors one by one.

6.4   Dark matter and normal matter

While discussing the density fluctuations (or 'density contrast'; we will use these terms interchangeably) in the last section, we did not bother to answer the question: density fluctuation of what? This question is relevant because the universe is made of many constituents. To begin with, the universe contains atomic nuclei made of protons and neutrons – which are usually called 'nucleonic matter' (of course, it also contains electrons, but electrons contribute very little to the mass compared to nucleons); it contains radiation made of photons; and it might very well contain large quantities of dark matter, possibly made of some other exotic variety of particles. (We saw in the last chapter that there is strong evidence to suggest that the universe contains vast quantities – nearly 10 to 100 times more – of dark matter compared to normal nucleonic matter.) In studying the density fluctuation, we need to be clear as to whether we are talking about density fluctuation in dark matter or nucleons or radiation. Let us look each one of them separately.

The simplest among the three happens to be density fluctuations in the radiation. Since 'normal' matter (made of protons, neutrons and electrons) was in a plasma state in the early universe, it was strongly coupled to the photons. Photons were constantly absorbed and emitted by protons and electrons in the early universe. This implies that any density variation in normal matter could produce a corresponding density variation in the radiation and vice versa. Roughly speaking, this arises because a region which is slightly overdense will emit and absorb more photons, thereby interacting with the radiation more strongly. Similarly, a region which is less dense than the average will interact with photons slightly weakly. This coupling between matter and radiation will ensure that the density fluctuation in these two components will be comparable. But one must also remember that radiation is made of photons which are travelling with the speed of light. It is therefore very difficult to make radiation cluster due to its own self-gravity. The speed of the photons helps them to escape the gravitational attraction quite effectively, and hence there is no (significant) clustering of radiant energy at any scale. Since normal matter and radiation are strongly coupled, it follows that one cannot have any significant clustering taking place in normal matter as well. While the self-gravity of matter tries to make it collapse, the radiation – which is strongly coupled to matter – tends to drag it away, preventing the collapse. Since the radiation is streaming at the speed of light, it wins over gravity, and matter is dragged away with the radiation. In short, there is no hope for the gravitational instability to operate (and density fluctuations to grow) in normal matter when it is strongly coupled to radiation.

It might seem that the wonderful picture we painted in the last section regarding the formation of structures has been thoroughly spoiled by the above fact. If density fluctuations cannot grow, then how is it that we can form structures like galaxies, etc. which are made of normal matter? Actually, all hope is not lost. You must remember that the radiation can prevent the growth of density fluctuations in normal matter *only when it is strongly coupled* to the matter. We saw in the last chapter that when the universe was about 1000 times smaller, electrons and ions combined to form neutral atoms and the radiation ceased to interact with matter. After this moment, radiation and matter were decoupled and they evolved as independent entities. The radiation, of course, kept moving at the speed of light without any significant clustering. But the density fluctuations in matter could now grow due to self-gravity, since there was no coupling between the radiation and matter. We thus reach the important conclusion that density fluctuations in normal matter could have grown only after radiation and matter decoupled. The decoupling occured when the universe was 1000 times smaller or, in other words, when the redshift was $z = 1000$.

While that is good news, it turns out that it is not good enough. The difficulty arises from the rate at which density fluctuations can grow in an expanding universe. We saw that when matter and radiation are coupled strongly, the density fluctuations in matter cannot grow at all. After they have decoupled, it turns out that the density fluctuations can grow in proportion to the expansion. That is, if the universe expands by a factor of 10, the density fluctuations can grow by a factor of 10. How large should the density fluctuations be before one can form structures like galaxies, which are held together by self-gravity? Since the self-gravity is governed by the local, enhanced, density distribution, while the expansion of the universe – which is trying to pull the structure apart – is governed by the background density, it follows that the local density enhancement should not be small compared to the background density for the structures to form. Taking a very, very conservative stand, we need at least a density contrast of about 1 to 10 to form structures like galaxies. But since the universe has expanded only by a factor of 1000 from the moment of decoupling till today, the density contrast could have grown at most by a factor of 1000 during this time. To get a density contrast between 1 and 10 today, we need a density contrast of 0.001 to 0.01 at the time of decoupling, give or take a factor of 3, say. So just before the matter and radiation decoupled, the density fluctuations in the universe must have been around this value for *both* radiation and matter. (Remember that when matter and radiation were coupled together they would have had comparable density contrast.) Any density fluctuation in the radiation would manifest itself as a corresponding fluctuation in the temperature. If the density fluctuation was between 0.001 and 0.01, we would expect the temperature of the radiation to show a fractional change of this amount in different directions in the sky. This is a definite prediction which we can make from the theory.

Unfortunately, observations invalidate this prediction. We saw in section 6.2 that the temperature fluctuation seen in the microwave background radiation is about one part in 100 000, which is smaller by a factor of 100 or so compared to the above prediction. In other words, the density fluctuations in the radiation (and in the matter) is about a factor of 100 smaller than what we need to form galaxies at the present epoch. This is the first example of how high quality observations can constrain models for structure formation. While a model based on normal matter and radiation (with the density fluctuations in matter growing after the decoupling) would have been perfectly valid as a theoretical scenario for structure formation, such a model is ruled out by direct observations.

Fortunately, there is still a way out. Note that the observations which ruled out the above scenario were based on the density fluctuation of radiation. Because normal matter (made of charged particles like protons and electrons) was coupled to radiation at redshifts $z > 1000$, they both shared the same density fluctuations. So any observational constraint on radiation will imply an indirect constraint on the density fluctuations of normal matter. But the situation is quite different as regards the dark matter. Dark matter is made of neutral, uncharged particles and does not couple directly to the radiation; so any constraint on the radiation will not constrain the density fluctuation in dark matter. This allows us to construct a more complicated, but potentially viable, scenario for structure formation along the following lines.

Let us assume that, at very high redshifts, there existed a certain amount of density fluctuation in dark matter and in normal matter. Since normal matter is coupled to radiation, they will have comparable density fluctuations. As the universe evolves, the density fluctuations in dark matter will grow in proportion to the expansion, while the density fluctuations in normal matter will grow very little (because it is strongly coupled to the radiation). At $z = 1000$ the density fluctuation in dark matter could be larger than that in normal matter (or radiation) without violating any observational constraint. After normal matter has decoupled from radiation, it will come under the gravitational effect of dark matter. Very soon, because of the dominant gravitational influence of the dark matter, the density contrast in normal matter will increase and they will share the density fluctuations of the dark matter. Subsequently, density fluctuations in dark matter and normal matter will continue to grow, eventually forming structures like galaxies.

To see this process in detail, we must note a key difference between dark matter and nucleons: dark matter particles have no electric charge and hence they are not directly coupled to the radiation. For this reason, a bunch of dark matter particles held together by self-gravity cannot lose their energy. The size of such a halo will not change significantly once the self-gravity starts playing a crucial role. Such a (dark matter) halo will act as what is called a 'potential well'. (This usage arises from the fact that such a halo will attract particles around it due to gravity, just as particles will tend to fall into a well.) A universe populated by dark matter halos of different sizes and shapes can be compared to the surface of a floor with potholes of different sizes. If we pour water on such a surface, it will collect in the potholes with more water gathering in deeper holes. Something similar happens to nucleons in a universe punctuated with dark matter potential wells. Nucleons fall into 'potholes' made of gravitational potential wells of dark matter halos. But you must remember that nucleons are made of charged particles, and hence can radiate away their energy. In other words, a hot gas of nucleons can cool by emitting radiation. As the gas cools, its temperature and hence its kinetic energy decreases, and it will move towards the lowest point in the potential well. Thus nucleonic matter will end up in the centres of dark matter halos.

The idea described above forms the cornerstone of several models of structure formation and is believed by most of the working cosmologists. It is true that the model relies heavily on the fact that density fluctuations of dark matter will help the growth of gravitational instability. However, you must remember that there is independent evidence for the existence of dark matter in galaxies and clusters. Nobody invented dark matter just to make up the above scenario. In fact, since observations of galaxies and clusters show that they are surrounded by vast quantities of dark matter, it is mandatory for us to worry about the density fluctuations in the dark matter as well. As a bonus, such a study helps us to solve the problem faced by a universe made of just normal matter and radiation.

The above picture, though basically correct, is highly oversimplified; and we have cheated a little in the description. To begin with, we said that the density fluctuations grow in proportion to the expansion of the universe; that is, if the universe expands by a factor of 10, the density fluctuations grow by the same factor of 10. This is not quite true. First of all, it is true only when the density fluctuations are small compared to unity. Once a region of the universe becomes very overdense, its behaviour will essentially be governed by self-gravity rather than the expansion of the universe, and the above result no longer holds. Secondly, this relation is not valid in the early universe when the universe is dominated by radiation. Such a universe expands very rapidly and does not allow the density fluctuations to grow by any significant amount. Finally, this law is true only in a universe with a mean density equal to the critical density. If the universe has less than critical density, then the growth of density fluctuations is slower. But it turns out that none of these facts really affect the conclusions we have arrived at (which is why we did not mention these complications earlier!)

Another 'cheating' had to do with the coupling between radiation and dark matter. The above discussion tacitly assumes that dark matter and radiation are completely uncoupled from each other. Well, this is also not *quite* true. If the dark matter has density fluctuations, these density fluctuations will produce variations in the gravitational force around them. The photons making up the radiation field do feel this variation in the gravitational force and are affected by it. Thus you see that the fluctuations in dark matter can *indirectly* affect the radiation field through the gravitational force. But in the case of normal matter, the coupling is a lot stronger. When normal matter was in the plasma state, it was made of ions and electrons which scatter radiation strongly. Compared to this, we may say that the coupling between dark matter and radiation is quite weak. It is because of this that we ignored the coupling between dark matter and radiation in the above discussion.

The fact that the dark matter is coupled to the radiation, albeit weakly, has an important consequence. We saw earlier that whenever two components are coupled, there is a tendency for the density fluctuations in one component to induce density fluctuations in the other. In the case of normal matter and radiation, this coupling is fairly strong, and hence we can put constraints on the density fluctuations in normal matter from observations related to radiation. In the case of dark matter, the coupling is indirect, but it is still possible to use observations related to radiation to put some constraints on the density fluctuations of dark matter. So we cannot have complete freedom in choosing the behaviour of dark matter just to suit our convenience. In fact, you will see that the existing observations do constrain the nature of dark matter fairly significantly.

6.5   Dark matter in two flavours: hot and cold

We saw in the last section that the driving force behind the formation of structures is the density fluctuations in the dark matter. Clearly, the observed structures in the present day universe will be related to the kind of fluctuations in the dark matter. We shall now take a closer look at the pattern of these fluctuations.

Let us start with a very early phase in the universe, when the universe was, say, a hundred million times smaller than it is now. We saw in the last chapter that the temperature of the universe scales inversely as the expansion; so when the universe was a hundred million years smaller, it also would have been a hundred million times hotter and will have had a temperature of about 300 million kelvin. The matter would have been in the form of a hot soup of plasma. By and large, this matter would have been distributed uniformly in space; but we do expect tiny fluctuations in the density. (These fluctuations must have been created at some still earlier epoch; we shall consider mechanisms for the generation of such fluctuations in a later chapter.) As we saw before, the fluctuations can be characterized by giving the density fluctuations at different scales which were present at this epoch. The key question is: what is the pattern of these density fluctuations?

Astronomers have classified the dark matter into two broad categories called 'hot' and 'cold'. These two kinds of dark matter lead to very different forms of density fluctuations in the universe and, consequently, to very different kinds of evolutionary history for the formation of structures.

To begin with, one needs to understand the adjective 'hot' (and 'cold') when applied to dark matter. As we said earlier, there is considerable evidence to suggest that the dark matter is made of some exotic variety of elementary particle. Since dark matter is assumed to be electrically neutral, the key characteristic of this elementary particle will be its mass. Let us suppose for a moment that its mass is very small compared to all other elementary particles known to us. The lightest particle known to us is an electron (see chapter 2) which has a mass of about 0.5 MeV. (Of course, we are only considering particles with nonzero mass; a photon, for example, has zero mass.) So let us assume that the dark matter particle has a mass, of say, 10 eV, which is 50 000 times lighter than the electron. In the very early universe, all the particles would have been colliding and interacting with each other and would have shared a common temperature. We saw in chapter 2 that the temperature of a system is essentially a measure of the kinetic energy of the individual particles. Therefore, in the very early universe, both the dark matter particles and the electrons must have had the same kinetic energy. But since the dark matter particle was far lighter than the electron, it had to move with much higher speed in order to have the same kinetic energy. As the universe cooled, there will have come an epoch when the dark matter particle effectively decoupled from the rest of the matter. But, of course, it would still be moving with very high speeds since it was very light. The adjective 'hot' merely means that the dark matter particle was moving with very high speeds (almost comparable to that of light) when it decoupled from the rest of matter.

On the other hand, if the dark matter particle was much heavier than the familiar elementary particles, the situation would be reversed. The proton, for example, has a mass of about 1 GeV. Suppose we have a dark matter particle with a mass of, say, 50 GeV. By the same arguments as above, we conclude that such a dark matter particle will be moving very slowly compared to protons and electrons in the early universe. So, it will have had a very low speed when it decoupled from the rest of the universe. Such a dark matter particle is called 'cold' dark matter.

The pattern of density fluctuations which would have existed in the dark matter during the early phases of the universe depends on whether the dark matter was hot or cold. Let us first consider the case of hot dark matter which is made of particles moving very rapidly in the early universe. Suppose that, at some given moment in time, we distributed the hot dark matter particles so as to produce some density fluctuation. (See the top frame of figure 6.2). Regions marked A and C have more hot dark matter particles (than the average) while the region marked B has fewer. Under normal circumstances, we would have expected the density in a region like A or C to grow due to self-gravity. In fact, this was the key idea introduced in section 6.1 (see figure 6.1). But when we described the gravitational instability in section 6.1, we tacitly assumed that the particles do not have significant speeds. This assumption is badly violated in the case of hot dark matter. After all, hot dark matter is 'hot' because the particles are moving with large speeds. Because of this motion there will be a tendency for particles in regions A and C to stream into region B, thereby reducing the density at A and C and increasing the density at B. (Of course, some of the particles in A will also move to the left and some of the particles in C will also move to the right; we have not bothered to show that in the figure.) You will notice that this has exactly the opposite effect to gravity; now the rapid motion of particles is wiping out the density fluctuation which was originally present!

Does it mean *no* density fluctuations can survive in a universe with hot dark matter? No. That is not true. To see why, consider the bottom frame of figure 6.2, which shows a density fluctuation with a much longer wavelength. Now the regions A, B and C are located quite far apart. The hot particles will certainly move from the peaks A and C to the nearby troughs like B, but since the distances between these points are much larger than the typical distance which can be covered by the rapidly moving particles, the fluctuations cannot be completely wiped out. We see that very large-wavelength fluctuations can survive in a hot dark matter universe, while fluctuations at small scales will be wiped out.

What determines the scale below which fluctuations cannot survive? As you can see from figure 6.2, this is essentially decided by the speed of the hot dark matter particle which, in turn, is related to its mass. The higher the mass, the

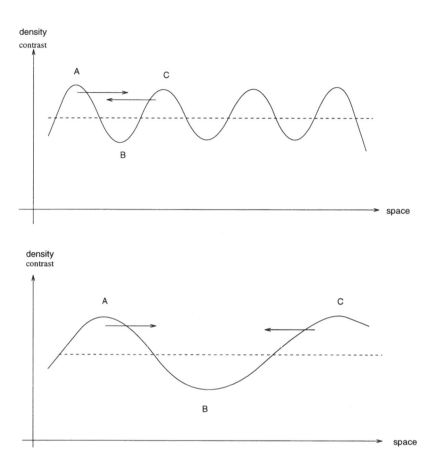

Figure 6.2 Small-scale density fluctuations cannot be sustained in dark matter if the particles are moving too fast. The rapid motion of the particles from crests like A and C towards troughs like B will tend to equalize the density of the dark matter. (See top frame.) But if the wavelength of the density fluctuation is larger than the typical distance moved by the dark matter particles (see bottom frame) then this process cannot be effective. For this reason, small-scale fluctuations in density are wiped out in dark matter, while large-scale fluctuations survive.

smaller the speed will be, and the smaller the scale below which the fluctuations can be wiped out.

To proceed any further we need to know the mass of the individual dark matter particle. Unfortunately, we have no direct handle on this important parameter. One possible choice for hot dark matter will be the neutrino, provided it has a nonzero mass. People have tried very hard to measure the masses of the neutrinos in the laboratory. We know for sure that the masses of the electron neutrino, muon neutrino and tau neutrino are *less* than about 12 eV, 250 keV and 35 MeV respectively. All these neutrinos may have zero mass; if, on the other hand, we assume that each neutrino has a tiny mass of, say, 10 eV, then it will make a possible candidate for dark matter. Alternatively, we may assume that the mass of the muon neutrino is about 30 eV and the other neutrinos are massless. Such a value for the mass will account for the dark matter seen in galaxies and clusters and will be enough to provide the critical density to the universe. If we assume this value of the mass as a working hypothesis, then we can compute the scale below which the density fluctuations will be wiped out in a hot dark matter model. This scale turns out to be the one corresponding to large clusters. Thus, hot dark matter models will have very few density fluctuations at scales smaller than that of a cluster.

What about density fluctuations at very large scales? These have to be small because we know that the universe is quite smooth at large scales. So we expect the density fluctuations in hot dark matter to be small at both small and large scales, and significant only in the intermediate scales. Figure 6.3 shows such a pattern of density fluctuations (see the thin line, marked 'HDM'). There are very few density fluctuations at scales smaller than that of clusters or larger than superclusters. The exact shape of this curve can be determined if we know the mass of the hot dark matter particle.

Let us next consider the cold dark matter (CDM) scenario, which turns out to be simpler in many respects. Since the mass of the CDM particles is so high, they move very slowly and are ineffective in wiping out fluctuations at any astrophysically relevant scales. So the shape of the density fluctuation is like the one shown by the thick line (marked 'CDM') in figure 6.3. Thus, cold dark matter scenarios can have density fluctuations at all scales from that of dwarf galaxies to superclusters.

Figure 6.3 nicely contrasts the pattern of density fluctuations in cold dark matter models with that in hot dark matter models. (The thick line corresponds to CDM and the thin line corresponds to HDM.) In both the models we expect the universe to be smooth at very large scales; so the density contrast decreases with scale at large scales. In the HDM model, all the small–scale fluctuations would have been wiped out due to the rapid motion of the particles, and hence HDM models will contain very few density fluctuations at small scales as well. Clearly, HDM will have maximum density fluctuations at some intermediate scale. In contrast, CDM will have density fluctuations at all scales.

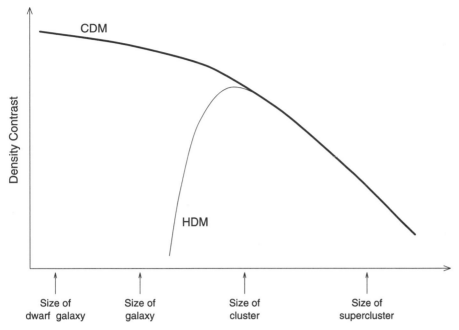

Figure 6.3 This figure compares the density contrast in the CDM and HDM models. In both models the density contrast decreases with increasing scale at large scales. This is because the universe is smooth at large scales in any model. The rapid motion of HDM particles wipes out density fluctuation at small scales. So the HDM model has maximum density fluctuation at some intermediate scale with decreasing density contrast at both smaller and larger scales. In the CDM model, the motion of the particles is so slow that it cannot wipe out the fluctuations at any astrophysically interesting scales. Hence CDM has a lot more density fluctuations at small scales compared to HDM.

The curves in figure 6.3 are based on purely theoretical considerations. Given such a pattern of fluctuation in some early epoch in the universe, we can calculate the nature and distribution of various structures which can form as the universe evolves. By comparing the result of such a theoretical calculation with observed properties of structures, we can either verify or reject the pattern of density fluctuations shown in figure 6.3. By and large, this is the game of structure formation played by cosmologists. The basic idea is to come up with the correct pattern of fluctuations which is consistent with all observations related to the structures which we see in the universe. As we shall see, it turns out to be rather difficult to play this game successfully.

6.6 Why hot dark matter does not work

Let us now consider the formation of structures in a universe with hot dark matter, made of massive neutrinos. This was the first model with dark matter which was analyzed seriously by cosmologists; unfortunately, the model does not work. That is, it leads to a universe which is quite different from what is observed.

The universe in this model is built out of several ingredients. First of all, we need to know the mass of the hot dark matter (HDM, for short) particle and its abundance. Based on the estimates of dark matter present in galaxies and clusters and from certain theoretical considerations, one can have a good guess at these parameters. It is usual to take the mass of the HDM particle – say, the muon neutrino – to be about 30 eV and its abundance to be about 330 particles per cubic cm. This mass of 30 eV makes it considerably lighter than the electron, which has a mass of 0.5 MeV. But the abundance of neutrinos is much higher than that of electrons. (There is only about 1 electron per cubic *meter* of the universe.) The high abundance more than compensates for the low mass and, in this scenario, the universe will have nearly 10 times more dark matter than visible matter.

The second ingredient in such a model is the shape of the density fluctuation which was present in the very early universe, say, at a redshift of $z = 1000$. We have discussed this feature in the last section; the thin line in figure 6.3 describes the HDM model. There is very little density fluctuation at small scales or large scales, and the maximum fluctuation is at a scale corresponding to large clusters of galaxies. This *shape*, it turns out, can be predicted from purely theoretical considerations, once we know the mass of the neutrino. What cannot be predicted from theory is the *actual value* of the density perturbations. That is, we do not know from theory whether the maximum value of the density contrast is 0.001 or 0.01 or 0.5 . . . This quantity is called the 'amplitude' of the density fluctuations. (If we know the value of the density fluctuation at *any* scale, then, knowing the shape, we can fix it at all other scales.) If we do not know the amplitude, then we have to treat it as an unknown parameter in the theory, and construct models with different values for this amplitude and compare them with observations. If we are lucky, some value of the amplitude will lead to predictions which agree with observations and we will be through. (Actually, it turns out that there is a way of fixing this amplitude from recent observations of microwave radiation; we will discuss that in detail later. For the moment, let us pretend that we do not know what this quantity is.)

The third ingredient in the model is, of course, the rate at which the fluctuations grow from $z = 1000$ till today. As we said before, we may assume that (for most of the redshift range) the fluctuations grow in proportion with the expansion. That is, when the universe expands by a factor of 10, the amplitude of the fluctuations increases by a factor of 10. This increase occurs at all scales uniformly, and hence the shape of the fluctuations does not change with time. Figure 6.4 shows the result of such an evolution. The top curve is the fluctuation spectrum at a redshift of 100, while the bottom one is the original fluctuation spectrum at a redshift of 1000. Between these two epochs, the universe has expanded by (about) a factor of 10, and the amplitude of the fluctuation has increased by the same factor. Each of the points in the top curve

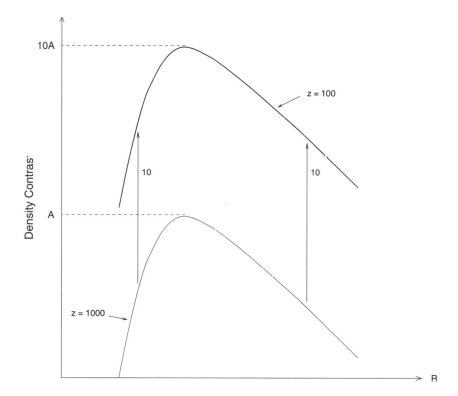

Figure 6.4   As the universe expands, the density contrast in HDM grows in proportion to the expansion. The figure compares the density contrast at a redshift of $z = 1000$ and $z = 100$. The universe has expanded by a factor of 10 between these redshifts, and the density contrast has increased by the same factor at all scales. The upper curve (at $z = 100$) is obtained from the lower curve (at $z = 1000$) by scaling up the amplitude everywhere.

is obtained by moving the corresponding point in the bottom curve vertically up by a factor of 10. The peak, which was originally at some value A, has now moved to a value which is 10 times as large, that is, to $10A$. (That is, if the original amplitude was, say, $A = 0.001$, then the amplitude would have increased to $10A = 10 \times 0.001 = 0.01$.)

This type of evolution leads to a very important conclusion. We saw earlier that in order to form structures like galaxies, clusters, etc., which are held together by their own self-gravity, the density contrast has to reach a value of about unity. As time goes on, the density contrast at all scales is increasing at the same rate; so clearly, the peak scale will reach a value of unity first. For example, suppose that the peak amplitude was 0.001 (that is, $A = 0.001$) at $z = 1000$. When the universe expands by a factor of 1000, this peak value will have reached $0.001 \times 1000 = 1$. At all other scales, the amplitude will be still less than unity. This shows that the first structures which form in a HDM universe will correspond to the scale at which the fluctuations are maximum. This is, in turn, decided entirely by the mass of the neutrino. For a mass of about 30 eV, the structures which form first will be large clusters containing about $4 \times 10^{15} M_{\odot}$ in dark matter.

To proceed further, we need to know how the density fluctuations evolve after the formation of the first structures. Here, unfortunately, all kinds of complications come up. The simple rule which we gave – viz. fluctuations grow by the same factor as the universe expands – is valid only when the amplitude of the fluctuations is small compared to unity. When the peak scale has an amplitude which is comparable to unity, it is difficult to analyse theoretically the further evolution of structures that have formed. To study such an evolution, it is best to simulate the dynamics of the universe using large computers. In such computer simulations one can evolve the universe further, and finally predict what kind of structures will exist today. Such simulations show that in a universe with the neutrino as dark matter, structures like galaxies have to form by fragmentation of larger, cluster-like, objects. What is more, these galaxies tend to stay close to each other because they have all originated from similar parent clusters. In the actual universe we do see some clustering of galaxies, but not as much clustering as predicted by the neutrino model. Thus we have to discard this model as not being in agreement with observations.

The above conclusion was reached by several cosmologists in the early 1980s, and the neutrino model was discarded as a viable candidate for structure formation. At that time, there was no reliable observation available regarding the amplitude of the density fluctuations; cosmologists only knew the overall shape. It was found that the model will not work *whatever* the amplitude. More recently, observations of microwave radiation have allowed us to fix the amplitude of the density fluctuation as well. (We shall discuss these observations in detail later.) These observations show that at $z = 1000$ the peak amplitude is about 4×10^{-4}. This fact spells yet another disaster for the HDM model which is far more serious. Note that from the redshift of 1000 till today the universe has expanded only by a factor of 1000. So the peak amplitude could have grown only up to a value of $1000 \times 0.0004 = 0.4$ till today. In other words, the peak value has not even reached the value of unity yet. But to form structures like galaxies, clusters, etc., it is absolutely essential that this amplitude has a value of at least unity. The neutrino model, therefore, cannot account for any of the structures which we see around us! If this model is true, structures will form only at some time in the future. For these reasons, HDM models lost their popularity and the attention shifted to the CDM models.

6.7   Scenarios with cold dark matter

Let us next ask how the cold dark matter scenarios fare in producing structures in the universe. In this case, we must remember that the initial density fluctuations in the universe are given by the thick line in figure 6.3. There are a lot of density fluctuations at small scales, with fluctuations decreasing gently up to the scales of galaxies and more rapidly at larger scales. Suppose we start with such a pattern of fluctuations at a redshift of say, 1000; what kind of universe will result at present?

To make any concrete prediction, we again need to know the amplitude of the fluctuations. Let us pretend once again that we do not know this quantity and treat it as a unknown number. As the universe evolves, the fluctuations grow at the same rate as the expansion of the universe, and the situation will be as shown in figure 6.5. When the universe evolves from $z = 1000$ to $z = 100$, the fluctuations at all scales will go up by a factor of 10. Since the CDM fluctuations in figure 6.3 decrease steadily as we go to larger scales, it is obvious that the fluctuations at small scales will reach the value of unity first. We saw earlier that when the fluctuations at any given scale reach the value of unity, one can form objects of that size held together by self-gravity. So, in the cold dark matter scenario, the first objects to form will be small-scale structures with sizes even smaller than galaxies.

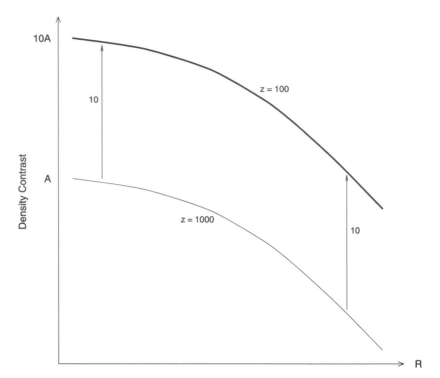

Figure 6.5   The density contrast in CDM models grows along the same lines as that of HDM models shown in the previous figure. When the universe expands by a factor of 10, the amplitude of fluctuations increases by a factor of 10 at all scales.

To evolve the universe further, we need to study how density fluctuations grow when they become comparable to unity. As we said before in the context of hot dark matter, this is a tricky affair and cannot be dealt with analytically. The best way to proceed is to simulate the universe in a computer and study the evolution of structures further. Such a study shows that in the case of cold dark matter scenarios, the formation of structures proceeds from small scales to large scales. As time goes on, larger and larger structures are produced in the universe, partly due to such objects forming ab initio and partly due to the merger of smaller objects. Such a progression is called 'hierarchical clustering' or, more expressively, as the 'bottom-up scenario'.

You must have noticed that the cold dark matter and hot dark matter models lead to very different evolution, mainly because of the difference in the shapes of the fluctuations shown in figure 6.3. Hot dark matter has very little power at small scales and has maximum fluctuations at scales corresponding to large clusters. As the universe evolves, it is the fluctuations in this scale which hit unity first; therefore, the first structures to form are large clusters. Galaxies etc. have to form later on due to the fragmentation of larger structures. This scenario is one that proceeds from the 'top-down'. In contrast, CDM scenarios produce sub-galactic size objects at first, and the evolution proceeds from the 'bottom-up'.

Can CDM scenarios reproduce the structures which are seen in the universe? The answer is quite complicated and depends crucially on the amplitude of the fluctuations. When this scenario was studied in detail, it was found that one ended up with too much clustering of galaxies, if one assumed that every dark matter halo contained a galaxy. Such a model is therefore ruled out by observations. That is, you cannot identify each dark matter halo with a galaxy in this scenario. The model, however, could be made to work, provided galaxies form only inside the halos which contain a sufficiently large amount of gravitational potential energy. This idea, called 'biasing', requires some explanation.

In our descriptions of structure formation so far, we have concentrated entirely on the dark matter part. This is, of course, due to the fact that there is nearly 10 to 100 times more dark matter in the universe than normal nucleonic matter, and hence the dynamics is basically governed by the evolution of dark matter. But eventually we have to compare the predictions of the theory with the observed properties of galaxies made of visible matter. Since observations suggest that visible matter is embedded in large dark matter halos, we need to understand how nucleonic structures form within the dark matter halos.

We saw earlier that nucleonic gas can cool and sink to the bottom of the potential wells formed by the dark matter halos in the course of evolution. This process could be more effective if the dark matter potential well is deeper. It is therefore possible that galaxies which we see actually materialize only in dark matter halos which are sufficiently deep, and not in *all* the potential wells. In other words, we should associate galaxies only with deep potential wells – not with all the potential wells – thereby introducing a 'bias' between visible matter and dark matter. It turns out that if we allow for this possibility, then we can make the CDM model work by adjusting the amplitude suitably.

In such a scenario the amplitude of the CDM fluctuations will be about 0.5 at a size of 16 Mpc. At larger scales, the amplitude will be less than 0.5, and at smaller scales the amplitude will be larger. Since galaxies form only at the deepest potential wells, the fluctuations in the galaxies can be about a factor of 2 larger. Hence the fluctuation in visible matter in the present day universe will be about unity at 16 Mpc; it will be smaller at larger scale and vice versa. Observations show that this is indeed true: the fluctuation in visible matter computed from galaxy surveys is about unity at 16 Mpc.

This wonderful picture, unfortunately, turns out to be again wrong. It turns out to be wrong in the noblest traditions of science: the scenario made concrete predictions about different phenomena in the universe which – when tested – were found to be not true.

To see how this comes about, one has to remember that the shape of the CDM fluctuations given in figure 6.3 is fairly certain. What was uncertain was the amplitude. We saw above that in order to reproduce the current universe the amplitude has to be about 0.5 at 16 Mpc. Given the amplitude at any one scale and the shape of the fluctuations, we can calculate the amplitude at any other scale. In other words, we make predictions about the universe at every other scale which can then be tested against observations. To understand how the CDM model failed, we need to know what these predictions are and how they are compared with observations.

6.8   COBE: first glimpse of the truth

In section 6.2 we saw that the models for structure formation can be tested by comparing the predictions of this model with observations of microwave radiation. Since this comparison plays a vital role in deciding the fate of CDM models, let us first take a closer look at this phenomenon.

If you have been following the arguments closely, you will wonder how observations of microwave radiation can say anything about *dark* matter. After all, dark matter is uncharged and is not directly coupled to radiation; in fact, this was a major advantage of dark matter models over and above models which contained only normal matter.

The connection between dark matter and microwave radiation arises in a somewhat subtle fashion. It is indeed true that dark matter is not directly coupled to radiation. However, dark matter produces a gravitational field and this gravitational field can affect radiation. Once again this effect arises because of the equivalence between matter and energy. Radiation has energy, and hence will not only contribute to gravitational force (after all, in the radiation dominated universe it is this energy which is driving the expansion of the universe) but will also be affected by the gravitational force.

The simplest illustration of this phenomenon is as follows. Consider a beam of light sent up from some point on the surface of the Earth and received at a point located at some height (see figure 6.6). To go from point A to point B the light has to climb against gravity. In this process, it will lose some of its energy, just as a ball thrown up from a point A will lose some of its energy when it reaches B. The key difference between a ball and a light beam is the following: the energy of the ball is related to its speed, and hence, when a ball loses energy, its *speed* decreases. The energy of a light beam or a photon is related to its *frequency*, and hence its frequency will decrease when it goes from A to B. (The speed of light, of course, does not change in the process.) When the

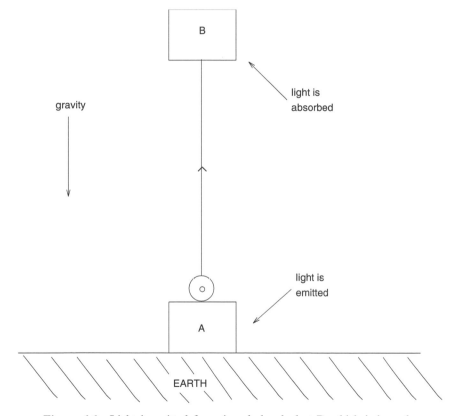

Figure 6.6 Light is emitted from A and absorbed at B, which is located at some distance vertically above A. During this process, the photons making up the light beam have to climb against the gravitational field of the Earth and thus will lose some energy. Since the energy is inversely proportional to the wavelength of the light, the light will appear redder at B. This process is called gravitational redshift.

frequency decreases the wavelength will increase and – using our familiar terminology – we can say that the light will be redshifted. This process, called the 'gravitational redshift' has been actually verified in the laboratory and is theoretically quite well understood.

We are now in a position to understand how dark matter can disturb the microwave radiation. Remember that the microwave radiation which we are detecting today is coming to us straight from a redshift of $z = 1000$. Let us suppose that at that redshift the dark matter distribution had some density fluctuations. Regions with higher dark matter density will be exerting larger gravitational force on the surroundings; we may say that regions with higher dark matter density produce larger gravitational potential wells in the universe. The microwave photons which are coming to us have to climb out of these potential wells. Those photons which came to us from deeper potential wells would have lost more energy compared to those which came from relatively shallow potential wells. Thus, the density fluctuations in the dark matter can induce fluctuations in the energy density of the microwave radiation. Since the energy density of radiation is, in turn, related to the temperature of the radiation, this process will lead to fluctuations in the temperature of the microwave radiation. That is, the radiation which we receive from different directions in the sky will have a tiny difference in temperature. The angular size at which these temperature fluctuations will appear in the sky will depend on the spatial scale of the dark matter potential wells at $z = 1000$. Small-scale density fluctuations will lead to temperature fluctuations at small angular scales in the sky, and large-scale fluctuations will lead to fluctuations at large angular scales in the sky. The amount of temperature fluctuation, in turn, will be related to the amplitude of the density fluctuations – the key quantity we are after.

So you see that the density fluctuation in the dark matter leaves its signature in the angular pattern of temperature fluctuations in the microwave radiation. If we could measure the temperature fluctuations at different angular scales, we would be able to figure out both the amplitude and the shape of the density fluctuations at different scales in any given model. Thus, in principle, we will have the complete information which is needed to construct the present day universe. In real life, of course, things are a little bit more complicated. To begin with, the process described above – viz. the loss of energy of photons while climbing the gravitational potential well – is only one among many different processes which can lead to temperature fluctuations. In order to find the density fluctuations from the observed temperature fluctuations, we need to disentangle these different processes. Fortunately, this disentanglement turns out to be quite easy at large angular scales. At scales larger than about 10 degrees in the sky, the only significant contribution to temperature fluctuations arises from the effect described above. So the temperature fluctuations at large angular scales can be used to directly infer both the amplitude and the shape of the density fluctuation at large scales.

As we mentioned in section 6.2, the satellite COBE detected the temperature fluctuations in the microwave radiation in 1992. These measurements have now been analyzed by several cosmologists, to obtain information about the density fluctuations. The good news is that the shape of the fluctuations at large scales is consistent with those produced in the inflationary models, and is shown in figure 6.3. The bad news is that the amplitude of the temperature fluctuations is larger than that predicted by the CDM model described in the last section.

A better way of stating this result is the following. Since the temperature fluctuation of the microwave radiation (measured by COBE) is now firmly established, we must use that to fix the shape of the density fluctuations at large scales. Once this is done, the amplitude of fluctuations in a CDM scenario is fixed at *all other* scales. By evolving this model, one can calculate the amplitude at smaller scales in the present day universe. These calculations show that the amplitude of fluctuations at 16 Mpc is about unity. We saw earlier that observations require this number to be more like 0.5 or so. This may appear to be a small discrepancy, but it turns out to be vital. The pattern of galaxy distribution predicted in a CDM model will be inconsistent with observation if it has to be consistent with the COBE measurements. In brief, the CDM scenario is ruled out.

Notice that the CDM model has been ruled out because it could not reproduce consistently *two* separate observations at two different scales. We can have a CDM model which will produce the correct amplitude at 16 Mpc; but that will predict microwave temperature fluctuations which are about a factor of three lower than what is observed by COBE. On the other hand, we can have a CDM model which agrees with the COBE observations; then it will give too high an amplitude at 16 Mpc and lower scales. It is just not possible to take care of both these observations because the shape of the fluctuations in figure 6.3 is fixed. Once the amplitude is fixed at any one point, it is fixed everywhere; we have no further freedom left.

6.9 Desperate models

The CDM scenario became popular in the early 1980s and was strongly pushed as *the* model for structure formation by the mid–1980s. The first signs of trouble with this scenario surfaced in the late 1980s (based on some galaxy surveys which we shall discuss later) and the real death knell came in 1992 with the COBE observations. It was soon clear that CDM cannot make both ends meet.

You must have realized, from the discussion in the last section, that the key trouble with the CDM model is the shape of the density fluctuations. Suppose we choose the amplitude at large scales in such a way that it agrees with the COBE observations. Then the CDM model predicts too large an amplitude at smaller scales (say, at 16 Mpc, for example). Suppose we can tinker with this feature and somehow *decrease* the amplitude at small scales without affecting the large scales. If we could do that, then we might have a successful model.

To do this, we have to change the shape of the density fluctuations and so we are no longer talking about 'the CDM model'. We are thinking of a suitably modified model which has the same amplitude as the CDM model at large scales but smaller amplitude (than the CDM model) at small scales. We have to think up some model in which the growth of small–scale fluctuations can be slightly suppressed, relative to the CDM model. How can we do that?

To see how this can be achieved, we must remember that the expansion of the universe works in a direction opposite to the formation of structures. You may think that if we could make the universe expand faster, then we might be able to enhance the adverse effect of expansion on structure formation, thereby slowing down the growth of density fluctuations. But note that to make the universe expand faster, we have to add more matter to it. But then we will also be increasing the local gravitational forces, and hence will accelerate formation of structures locally. These two effects work in opposite directions, and eventually the local effect will dominate. So what we need to do is to add more matter to the universe without actually contributing more to the density fluctuations. That is, we need to add matter (or energy) density which is smooth over a large region. If we could do that then we could decrease the rate of growth of smaller–scale density fluctuations embedded in that region.

It is therefore obvious that we need to add a different kind of matter (compared to CDM) to make the scenario work. This matter, which we add, should not contribute to the small–scale density fluctuations and should be smooth over such scales. If we could do that, then we could suppress the rate of growth of fluctuations at small scales without affecting larger scales. The amplitude at 16 Mpc can be brought down while being consistent with the COBE observations.

But we already know one particular kind of matter which has exactly this property! It is just our old friend – hot dark matter. We saw earlier that small-scale fluctuations in hot dark matter are naturally wiped out due to the fast motion of the HDM particles. So, for scales smaller than large clusters, HDM behaves like a smooth distribution which could contribute to expansion but not to the local gravity. All we need to do is to make a universe with some HDM and some CDM. The density fluctuations at very large scales in such a mixed dark matter model (MDM, for short) will be the same as that of CDM and can be adjusted to reproduce the COBE observations. Since we now have a smooth distribution of HDM particles at smaller scales, the growth of density fluctuations will be slower compared to that of pure CDM models at these scales. In other words, we can tinker with the amplitude at 16 Mpc without touching the COBE observations.

The density contrast for this model is shown in figure 6.7. The top line is the standard CDM density contrast with the amplitude adjusted to agree with COBE observations at large scales. Near 16 Mpc, the amplitude in this model is about unity which, as we saw, was too high. The lowest curve is the corresponding density contrast in a model containing a mixture of HDM and CDM. You can see that these models have the same behaviour at large scales; but at small scales the MDM has lower density contrast than CDM, helping us to solve the difficulty faced earlier. (We will discuss the middle curve a little later.)

You would have probably guessed that, in order to achieve this satisfactory state of affairs, one needs to mix CDM and HDM in the right proportions. This is indeed true. We need to have about 30 per cent of the critical density contributed by HDM and 70 per cent by CDM; there is some latitude in these numbers but not much. We have to assume that the dark matter halos around the galaxies are now made of two *different* kinds of dark matter particles: massive neutrinos making up the HDM component and some other exotic particles, much heavier than protons, making up the CDM component. What is more, they must exist in the right proportion.

Unfortunately, this model turns out to be not viable even after we do all these. The reason has to do with the abundance of structures at high redshifts, and we will discuss it in detail in the next chapter in section 7.5.

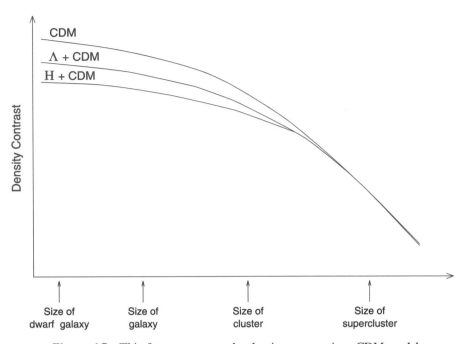

Figure 6.7 This figure compares the density contrast in a CDM model with two other models. The top curve is that of a CDM model with the amplitude adjusted so as to reproduce the COBE results. The bottom curve is that of a mixed dark matter model containing both hot and cold dark matter. The middle curve corresponds to a model with a 'cosmological constant'. These two (lower) models agree better with observations than the CDM model.

While the MDM model is a possibility, it is not the only possibility. As we said before, the key ingredient which we needed was a constant energy density smoothly distributed all over the universe. It is possible to do this in a quite different manner by introducing an entity called 'the cosmological constant'. Cosmological constant is the name given to a hypothetical form of matter which has negative pressure. We saw in section 5.4 that such an exotic kind of matter can maintain constant energy density even while the universe is expanding. If this form of matter comes to dominate the expansion of the universe, then the universe will expand very rapidly and forming the structures will be quite difficult. In section 5.4 we saw that such a rapid expansion, if it takes place in the very early universe, can have some useful consequences. It is the same kind of matter we are talking about now, but at a very different epoch. We are not concerned with the early universe but a very late stage in the evolution.

(You may wonder why it is called the 'cosmological constant'. The reason is purely historical. In the early days of relativity, Einstein introduced constant energy density – with negative pressure – in his equations with the hope of producing a static, non-expanding, universe. This term in his equations was called the 'cosmological constant'. Even though the original motivation for this term no longer exists, the terminology persists. Any exotic matter with negative pressure which retains constant energy density as the universe expands is called the 'cosmological constant', for this historical reason.)

By adding some limited amount of energy density in the form of such exotic matter, one can slow down the rate of growth of density fluctuations. The idea here is basically the same as that in adding HDM. By its very nature, the energy density of this form is quite smooth and will help us in reducing the amplitude at 16 Mpc. Once again, we need to add the right mixture: too much exotica will prevent structure formation completely, while too little will not help to solve the problems faced by the standard CDM. Detailed calculations show that we need to have about 65 per cent of matter in the form of a cosmological constant and 35 per cent of matter in CDM particles. In such a model, the dark matter around galaxies, clusters, etc. is made purely from CDM.

The density contrast for this model is also shown in figure 6.7. The middle curve corresponds to the model under discussion. We see that the amplitude is not as low in this model as in the MDM model. Nevertheless it is sufficiently low to solve the difficulty faced by the pure CDM model.

How does this model fare? It turns out that the age of the universe in models with a cosmological constant can be somewhat higher than in the previous models (like MDM or CDM) which we discussed. This can be useful if observations of older stars in the galaxy show that they have an age of, say, 14 billion years. On the negative side, this model faces difficulties in explaining several other observations which we will discuss in the next section. But none of these observations are yet firm enough to categorically rule out this model.

In summary, it is only fair to say that say that we do not yet know the correct model for structure formation. The simplest and most elegant models like the pure CDM model or the pure HDM model are ruled out by observations. More complicated models always have more adjustable parameters, and by fiddling with them one can make the models agree with *some* observations. The acid test for these models will be whether they can account for *all* observations. We shall now turn to this question.

6.10   The gamut of observations

In discussing models for galaxy formation, we used observations at only two scales: the first was the COBE observations at very large scales and the second was the amplitude of the density fluctuations at 16 Mpc. Of course, in the full theory, we need to explain the observations at very many different scales. As we shall see, these observations put a strangle-hold on the theoretical models.

To begin with, it must be noted that the microwave background radiation can offer us a wealth of information regarding the state of the universe at $z = 1000$. The measurements from the COBE satellite, in some sense, are only a tip of the iceberg. By its very design, COBE could only sample the temperature fluctuations at large angular scale (about 10 degrees in the sky) and was insensitive to the temperature fluctuations at smaller scales. Measuring the temperature fluctuations at small angular separations in the sky will provide a direct handle on the density fluctuations at smaller scales. Several experiments are now in progress to measure the temperature fluctuations on these scales, and some of them have reported tentative detections. In a few years' time, the situation will become clearer and we will be seeing the pattern of density fluctuations in the sky at different length scales.

Comparing the temperature fluctuations at smaller scales – when detected – with theoretical models is, however, a somewhat complicated task. At large scales – like the 10 degree scale probed by COBE – the temperature fluctuations are essentially determined by the dark matter distribution. But at small scales, the primordial gas also distorts the temperature distribution and one has to estimate this effect accurately. Once this is done, we will have a handle on not only the dark matter distribution but also the gas distribution. What is more, different models predict different patterns of temperature fluctuations at small scales, even if they produce identical results at the large scales probed by COBE. We saw in the last section that two possibly viable models for structure formation could be the ones with either cold plus hot dark matter or with cold dark matter plus cosmological constant. Accurate measurement of small-scale temperature fluctuations will – in principle – allow one to distinguish between these two models.

There are several other observations which could supplement the measurements of temperature fluctuations. One of them is related to the motion of galaxies at very large scales. We have seen in chapter 5 that in an expanding universe galaxies are receding from each other with speeds which are proportional to their separation. This is true if the universe is completely smooth. But in the real universe there are large fluctuations in the density at scales smaller than about 100 Mpc. Hence, the galaxies which are separated by distances smaller than 100 Mpc will not move *strictly* in accordance with the Hubble law. For example, the Hubble law states that the speed of recession of a galaxy should be strictly proportional to its distance from us. However, in real life, this speed may be slightly different from what is computed by this relation. The Hubble law also demands that the galaxy must be moving directly away from us and should not veer sideways. Once again, if inhomogeneities are present, the galaxies may move slightly off-course. A drastic example of such a deviation is shown by the Andromeda galaxy which is at about 1 Mpc from us. According to Hubble's law, we would have expected it to be moving *away* from us with a speed of about 100 km s^{-1}. In reality, however, Andromeda is moving *towards* us with such a speed! The reason for this is, of course, obvious. In the local group – of which Andromeda and the Milky Way are members – there is a large excess density of matter causing a large, local, gravitational force. This local excess density makes Andromeda and the Milky Way move towards each other rather than recede from each other.

It is, therefore, clear that by observing the motion of a large collection of galaxies in a region and by checking whether they follow Hubble's law, we will be able to probe the local density in that region. If the local density is very different from the average density of the universe, then the galaxies in that region will also move very differently from the cosmic expansion. You must realize that this observation takes into account all the gravitational force which is exerted on the galaxies. If the region contains a large amount of dark matter in the form of a local density enhancement, its presence will be revealed through the errant motion of galaxies in that region. So we can actually measure the amount of dark matter by observing the motion of visible galaxies.

Before this idea can be put into practice, one needs to take care of two difficulties. Firstly, one needs to know exactly how our local group of galaxies is moving in the cosmos. Without knowing our own motion we will not be able to figure out the actual motion of other galaxies. Secondly, one must be able to measure the distance to the galaxies accurately by some procedure. The speeds of the galaxies can be measured directly from the Doppler effect in the spectrum. But this speed is partly due to the expansion of the universe caused by the uniform distribution of matter and partly due to deviations in density from uniformity. To measure the latter, we need to know the former, viz. the speed due to Hubble's law. Since this quantity is proportional to the distance to the galaxy, we need to know the distance to each of the galaxies.

The first of these difficulties can be handled fairly easily using observations of the microwave background radiation. Remember that the COBE satellite is not at rest with respect to the microwave radiation field: COBE is orbiting the Earth; the Earth is orbiting the Sun; the Sun is moving in our Milky Way galaxy and the galaxy itself is moving in space. All these motions, combined together, make the COBE move in some specific direction with respect to the cosmic radiation. Because of this motion, COBE will be hit by more microwave photons in front than at the back. (This is similar to what happens if you run in the rain; you will be hit by more rain drops in front than at the back.) From the COBE observations it is fairly easy to isolate this effect and thus determine the precise direction in which COBE is moving. Since we know precisely the orbital motion of COBE, Earth and the Sun, we can subtract out these effects and determine the direction and speed with which the Milky Way is moving in space. Cosmologists now know for sure where they are heading, thanks to COBE!

To tackle the second difficulty, astronomers have come up with fairly complex methods for determining these distances. Using these methods, they have been able to estimate the deviation in the speeds of the galaxies compared to the Hubble expansion. On the average, galaxies separated by distances of the order of 50 Mpc show a deviation of about 350 km s^{-1}. Note that two galaxies separated by 50 Mpc are expected to fly apart with a speed of 5000 km s^{-1}, due to the expansion of the universe caused by the uniform distribution of matter. The density fluctuations at a scale of 50 Mpc seem to be causing only a deviation of about 350 km s^{-1} or roughly about 7 per cent. This suggests that the non-uniformity of density at these scales is also roughly at a 7 per cent level. Given any theoretical model, one can compute precisely the amount of density fluctuation present at any scale, and hence the amount of deviation expected in the speeds of the galaxies. Comparing this theoretical expectation with observation, one can again check the theoretical model. For example, the model with the cosmological constant (discussed in the last section) does face some difficulty in explaining the large-scale motion of galaxies in the range of 10 to 100 Mpc. The velocities predicted in the model are somewhat lower than those observed. However, the observations are not yet firm enough for one to rule out this model on this count.

The observations discussed above – viz. those based on the microwave radiation and those based on large scale motion of galaxies – directly probe the distribution of all gravitating matter in the universe. There are other kinds of cosmological observations that are based on the distribution of *visible* galaxies. Galaxy surveys fall in this category. A galaxy survey is essentially a census of all galaxies in the universe, in which one tries to measure the angular position and the redshift of a large number of galaxies in the sky using the best available observational techniques. In practice, such galaxy surveys are severely limited by the availability of resources. To begin with, any given astronomical instrument (like a telescope) will be able to detect only objects which are brighter than a threshold value; the fainter the object is, the harder it will be to detect, and if it is too faint the instrument just will not detect it. For this reason, no survey can be thought of as really complete. Further, some galaxies may be faint in optical wave-bands but bright in infrared, say, or vice versa. Depending on the kind of instrument used to carry out a survey, we may pick up galaxies which are of particular kind, and could end up getting a distorted view of the universe. Finally, the measurement of galaxy redshift is fairly time-consuming, and hence one is limited by the availability of telescope time to carry out such systematic surveys. One must also add to these woes the fact that the astronomical instruments might not be able to scan the entire sky, and hence could only be able to probe selected regions of the cosmos.

In spite of all these difficulties, cosmologists have spent considerable time and energy in conducting galaxy surveys. As years progressed, these surveys became bigger and better and now give us a broad picture of the way galaxies are distributed in space. The most striking feature which has emerged from these studies is that the galaxies are distributed very, very, non-uniformly in the universe. Most of the galaxies seem to be located in sheet-like regions in the sky rather than appearing as isolated entities. Any theoretical model is expected to explain this feature of galaxy distribution.

Unfortunately, these observations have been the hardest ones to compare with theoretical models. The key difficulty is that theoretical models are good at predicting the distribution of dark matter in the universe but are not so good at predicting the distribution of visible galaxies. The behaviour of dark matter is governed entirely by gravitational forces, and hence is easy to incorporate theoretically – both in calculations and in computer simulations. To understand how visible galaxies are distributed, we need to study the cooling of gas and fragmentation, which are extremely difficult to handle theoretically. In the last few years several groups have attempted to develop computer simulations which will take the effects of gas dynamics into account. It is, however, only fair to say that these numerical simulations are still in their infancy, and we do not have the full picture of how the galaxies are distributed in space from theoretical calculations.

In the absence of such a detailed picture, astronomers have concentrated on some simpler aspects (related to the distribution of galaxies) in order to compare theory and observation. One of them is an indicator called the 'correlation function'. This quantity measures the amount of clustering in the distribution of galaxies. Suppose we consider a spherical region of radius 10 Mpc around any given galaxy. If all the galaxies were distributed uniformly in space, then one would expect around two galaxies within this volume. But if galaxies have a tendency to cluster together, then one would expect to find more galaxies within this volume. Observations suggest that, within a spherical region of radius 10 Mpc around any galaxy, one finds roughly twice as many galaxies as estimated for a uniform distribution. The correlation function is essentially the ratio between the excess number of galaxies and the mean number of galaxies in some region. Since a sphere of 10 Mpc actually contains four galaxies, while the mean number expected in this volume is two, the ratio of excess number of galaxies to mean number of galaxies is $(2/2) = 1$. So we say that the correlation function at 10 Mpc is about unity. You can play the same game by taking spheres of different sizes around each of the galaxies, and thus obtain the average correlation function for different scales. It turns out that the correlation function decreases roughly as the square of the size of the sphere increases. For example, the correlation function at 100 Mpc is about 0.01. Given the distribution of galaxies in space, based on some galaxy survey, one can determine this correlation

We expect theoretical models to predict this correlation function from first principles, so that we can compare theory and observations. The difficulty is, once again, the same as we mentioned before. Theoretical models are extremely good at predicting the correlation function for dark matter halos, but they are poor at predicting the correlation function for visible galaxies. If we assume that each dark matter halo contains a visible galaxy, then we can convert the theoretical predictions based on dark matter to observational results based on visible matter. Unfortunately, this assumption may not be true. In other words, there could be significant 'bias' between dark matter distribution and visible galaxies. This could arise, for example, if it happens that cooling of gas, etc. is favoured in deep potential wells but not in all potential wells. As we saw earlier, the original CDM models predicted the correct distribution for visible galaxies only when certain bias was introduced between the distribution of visible and dark matter.

In a full galaxy survey one tries to ascertain the exact location of the galaxy, which requires three bits of information. The first two are the angular position in the sky, which is fairly trivial to obtain. The third is related to how far the galaxy is from us, and can be determined only if the redshift to the galaxy is known. If we are willing to ignore the third bit of information and are content with just knowing the angular positions of the galaxies in the sky, then it is easier to make progress. If the galaxies are not distributed uniformly in space, then they will not be distributed uniformly in the sky either. If we photograph different regions of the sky and then scan these photographs to ascertain the galaxy positions, we will be able to determine precisely the sky distribution of the galaxies, and hence measure the amount of clustering in the sky. The idea here is similar to the one described in the last paragraph and involves determining an *angular* correlation function of galaxies. This task was completed in the mid 1980s using a large collection of photographic plates of the sky. The resulting angular correlation function could then be compared with theoretical models in order to test the validity of the latter. It is this comparison which suggested for the first time that the CDM model was in trouble. It turned out that the CDM model which was popular in the early 1980s predicted a lower angular correlation function than observed. This difficulty was noted long before the COBE observations came in. Of course, the COBE observations re-emphasized the same difficulty, viz. that CDM has fewer density fluctuations at large scales than is seen in the observations. Models which agree with the COBE observations (like the MDM model) also tend to predict the correct results for angular correlation functions of galaxies.

But one must realize that the COBE observations probe the dark matter, while the angular correlation function is based on visible galaxies. The agreement of models with both these observations suggests that, at sufficiently large scales, visible galaxies and dark matter are distributed in the same manner without any bias. These observations probe scales of about 50 Mpc and larger. We also have evidence, based on the spatial correlation function of galaxies, that there is some bias between the distribution of dark matter and visible matter at smaller scales. All these go to show that the formation of visible galaxies is an extremely complex process, and we need theoretical models which are capable of making correct predictions at different scales.

6.11   The first structures

The descriptions in the previous sections give a broad overview of how structures bound by self-gravity can arise, starting from small density fluctuations in the early universe. Any successful scenario for such galaxy formation has, of course, to tie up this stage with the features of the currently observed universe.

To do this, we have to study the further evolution of a system made of dark matter and normal matter bound by its own self-gravity. The dark matter, made of uncharged particles, will form a reasonably stable structure of definite size and mass. In the initial stages, the cloud of gas contained in it will be made of just hydrogen and helium, with about 76 per cent of the weight being contributed by hydrogen. When this gas cloud contracts under the gravitational force of the dark matter, parts of the gas will collide with each other and the temperature will increase, and eventually a fair amount of hydrogen can get ionized. Further evolution will depend on how efficiently the gas clouds can cool and form galaxy-like structures made of a collection of stars. We shall first describe a cooling process and fragmentation which could lead to the formation of stars; then we shall discuss how morphological differences like ellipticals and spirals can arise.

The free electrons and protons in the ionized gas can cool by two different processes. The first process is simple radiation of energy by the charged particles. Electrons and protons, while moving erratically in the gas cloud, can radiate photons, thereby losing their energy. Secondly, the electrons can also lose their energy indirectly by colliding with neutral hydrogen atoms. When an electron collides with a hydrogen atom, it can kick the electron in the hydrogen atom from its ground state to a higher energy state, thereby transferring some of its energy to the hydrogen atom. The excited electron will, of course, fall back to the ground state within a short time, radiating away the extra energy. In the process some of the kinetic energy which the original free electron possessed has been transfered to radiation.

These two processes help the gas cloud to cool down, but not by a large amount. It turns out that these processes act somewhat like a thermostat. If the gas cools too much, the free electrons will lose a lot of energy and will tend to get captured back by the protons, thereby forming neutral atoms again. This process reduces the rate of cooling and tends to keep the temperature up. Similarly, if the temperature increases too much, the collisional processes, etc. also increase, thereby increasing the rate of cooling; which, in turn, will cool the system more efficiently. Calculations show that these processes maintain the temperature at around 10 000 Kelvin.

Because the temperature does not rise, the pressure of the gas does not increase fast enough to prevent the gravitational forces from crushing the cloud. As a result, the gas cloud fragments into smaller pieces, so that within each small piece the pressure could balance gravity. (This works because gravitational forces increase with the size of the object; at a given temperature, the pressure can support a smaller object but not a larger one.) But this is a losing game. Soon the smaller fragments will go through the same process and the gravity will start dominating the thermal pressure in them. These fragments will soon break up into smaller pieces.

When can this avalanche stop? Remember that the loss of energy was possible only because the photons emitted by the gas cloud could escape from the system. If this does not happen and the photons are trapped inside, then the system cannot lose energy; pressure will build up and rapid fragmentation can be stopped. In other words, the gas clouds should become opaque (rather than transparent) to radiation. This will eventually happen when the density of the gas clouds is sufficiently high.

Once the pressure forces are comparable to self-gravity, the fragmentation stops and each piece enters a stage of very slow contraction. It radiates at a very reduced rate because of the increased opaqueness to photons. Such an object is called a 'protostar', and grows in mass as more matter is accreted from the surroundings. A protostar will, in general, be also spinning because of the residual gas motion. The matter which accretes to the protostar will form a flattened, spinning, 'protostellar disc' of gas around it. The slow gravitational contraction of the protostar will continue until the temperature is high enough for nuclear reactions to take place at the centre. And then, a star will be born! The dense gaseous disc around a protostar will undergo its own dynamics, and there is every likelihood that planetary systems orbiting the star can form out of it.

The above process could account for the formation of a collection of stars starting from a gaseous cloud. Depending on the distribution of these stars, such a collection could look like an elliptical or a spiral. This broad division of galaxies can arise as follows. When a rotating gas cloud contracts under its own weight, it will have a tendency to get flattened in the direction perpendicular to the axis of rotation. If the rate of star formation is slow compared to the time taken for the gas to collapse, most of the gas will end up in a disc perpendicular to the direction of rotation before star formation takes place. In such a case, most of the stars will form in the disc and the structure will end up as a disc galaxy. On the other hand, if the star formation rate is high, stars will form all over the place even before the gas can collapse to a disc. Once star formation has progressed significantly, each star will move in its own orbit in the common gravitational field of the system and further collapse will not take place. In this case, we will end up with a more spread-out distribution of stars like in an elliptical.

Another equivalent way of understanding this difference is the following. When a star gets formed, it retains the speed with which it is born and moves around in the gravitational field of other stars and gas in some specific orbit. If the star was formed before the gas collapsed to a disc, it will move in some random direction. On the other hand, when the gas collapses to form a disc, it will lose most of its speed perpendicular to the disc, and any star which gets formed *afterwards* will only have motion in the plane of the disc. Thus elliptical galaxies have stars moving in a fairly spread-out fashion, while disc galaxies have stars moving mostly in the plane.

We, therefore, conclude that the ellipticals formed their stars very efficiently in a dramatic burst during the initial collapse, while the spirals probably formed only ten per cent of their stars in the initial phase. Most of the stars in a spiral galaxy formed in a leisurely fashion spread over the past ten billion years after the gas had already formed the disc. As a consequence, most of the light in an elliptical is from older, redder stars, while the spirals shine mostly due to younger, bluer stars (see figure 6.8).

How can we understand the formation of larger structures like groups and clusters? Here we need to take into account yet another factor: mergers of galaxies. In today's universe, collisions and mergers between galaxies are rare events, but not so in the past. Since structures developed hierarchically, we would have expected single galaxies to form before large groups or clusters. Computer simulations of such systems have shown that the merging of several smaller sub-units can lead to the formation of nearly uniform-looking spherical clusters. Similarly, clouds with masses of 10^6 to 10^7 $M_\odot$ used to merge frequently when the universe was about a tenth of its present age, and formed individual massive galaxies. Since such merging cannot be 100 per cent efficient, one would expect to find the massive galaxies to be surrounded by several dwarf-like companions. This is precisely what we see in the local group, which is dominated by the Milky Way and Andromeda, each of which is surrounded by several dwarf galaxies.

As you can see, the details of the above processes can be extremely complex. There are several details in the above story which astrophysicists are still debating about. The overall picture, however, seems to be the one outlined above. Assuming that the above picture is correct, what can we say about the properties of stars which we see in our galaxy? In other words, are there some signatures which could be used to bolster our faith in the above scenario? It

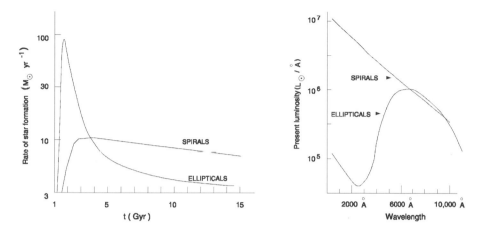

Figure 6.8   According to current models, the stars in the elliptical galaxies were formed in a rapid burst during a short span of time, while the star formation in spiral galaxies goes on at a reasonably steady rate over a long stretch of time (see the left frame). For this reason, most of the light from the ellipticals is due to older redder stars, while the light from spirals are due to younger bluer stars (see right frame).

turns out that there are some such signatures, though they are not completely free of astrophysical controversies.

One distinguishing characteristic of the initial bunch of stars formed along the above lines is the following. They formed out of a gaseous mix of hydrogen and helium without any other heavy elements being present. These stars will of course synthesize heavy elements (which are called 'metals' in astronomical jargon) in the course of their evolution. When the first generation of supernova explosions occurs, these heavy elements will be thrown out into the interstellar medium. Later stars which form in the gaseous interstellar medium will contain traces of heavy elements even in their initial composition. For example, the solar system – which is estimated to have formed about 4.6 billion years ago – has a fossilized record of the chemical state of the galaxy at that time. Nearly two per cent of the mass of the Sun is made of elements heavier than helium. In the Orion nebula – in which stars are forming at present (see figure 4.8) – the fraction of heavy elements is nearly twice as much as that in the Sun. This is because the interstellar medium at present has been enriched for an additional 4.6 billion years since the formation of the solar system. As a result, the stars which are forming today will contain nearly twice as big a fraction of heavy elements as the Sun. On the other hand, most of the stars in the galactic spheroid have less than about three per cent of the solar metal abundance. This clearly suggests that the stars in the spheroids must have formed earlier, before the gas settled into a disc. In fact the formation of the spheroidal component in a disc galaxy is similar to the formation of the elliptical galaxies. Both contain stars formed before the gas cloud has collapsed significantly.

There are also some systematic differences between the stars formed in different generations. For example, the electrons in the heavy elements (like carbon) are quite efficient in absorbing radiation. Hence, the second generation stars will be more opaque to radiation compared to the first generation stars which had no heavy elements. As a consequence, the first generation stars would have been more massive and brighter than, say, our Sun. They also would have been relatively short lived. We do find a surprising paucity of stars which are deficient in heavy elements. Even the oldest stars in our galaxy contain at least one per cent of the abundance of heavy elements found in the Sun. This is consistent with the idea that the first generation stars – which did not have heavy elements – were very short lived.

It is also seen that, in most galaxies, a stellar population with a higher abundance of heavy elements exists closer to the centre. The stars which are in the outer regions of the galaxy are relatively poor in abundance of heavy elements. Once again, this is what one would have expected in the above scenario. As the primary stars went through supernova explosions, they would have spewed out heavy elements, thereby producing an interstellar medium slightly richer in metal abundance. This enriched gas will fall towards the inner regions of the galaxy, and the second generation of stars will form closer to the centre. As they evolve, the process can get repeated along the same lines. The net effect will be to have stars with a relatively higher metal abundance near the centre and lower metal abundance towards the edge.

7

The universe at high redshift

7.1 Probing the past

The discussion in the last chapter shows that most of the prominent structures in the universe have formed rather recently. In terms of redshifts we may say that galaxy formation probably took place at $z < 10$. Our understanding of galaxy formation could be vastly improved if we could directly observe structures during their formative phases. Remember that in the case of stars we can directly probe every feature of a stellar life cycle from birth to death; this has helped us to understand stellar evolution quite well. Can we do the same as regards galaxies?

Unfortunately, this task turns out to be very difficult. The life span of a typical star — though large by human standards — is small compared to the age of the universe. This allows one to catch the stars at different stages of their evolution. For galaxies, the timescale is much longer and so we cannot hope to find clear signals for galaxies of different ages. Secondly, the distance scales involved in extragalactic astronomy are enormously large compared to stellar physics. This introduces several observational uncertainties into the study.

In spite of all these difficulties, astronomers have made significant progress in probing the universe during its earlier phases. We saw in chapter 5 that the farther an object is the higher its redshift will be. But since light takes a finite time to travel the distance between a given object and us, what we see today in a distant object is a fossilized record of the past. Consider, for example, a galaxy which is at a distance of one billion light years. Since light took one billion years to travel from that galaxy to us, a photograph of that galaxy reveals its condition one billion years ago. Thus, by probing distant objects we can also probe the condition of the universe at earlier epochs. The difficulty, of course, is that far away objects will appear fainter than nearby objects (of the same intrinsic luminosity) and hence will be more difficult to detect. With most of the optical telescopes available today, we can detect ordinary galaxies and clusters up to a redshift of about 1 or so, which corresponds to an epoch when the age of the universe was nearly 35 per cent of its present value. (The biggest telescopes are just beginning to detect galaxies at higher redshifts; but this field is still in its infancy.) This is not of much use, because most of the structure formation could have been completed by a redshift of $z = 1$. We need to probe objects which are still younger and have higher redshifts.

Fortunately for the astronomer, nature has supplied us with certain high luminosity objects which are visible even when they are very far away. Notable among them are the 'quasars' which are detected up to a redshift of nearly 5, and 'radio galaxies' which are seen up to a redshift of about 4.5. These objects allow us to probe the universe when its age was about 7 to 8 per cent of the current age (see figure 7.1).

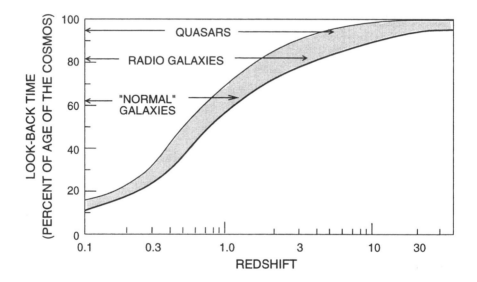

Figure 7.1   High redshift objects allow us to probe the universe when it was considerably younger. The typical age of the universe (as a fraction of the current age) is shown in the figure for different redshifts, as a shaded band. The actual age lies within this band and depends on the value of cosmological parameters. Quasars (see section 7.2) and radio galaxies (see section 7.3) allow us to probe the young universe.

7.2 Quasars: the enigmatic machines

The most familiar objects which we come across in astronomy are stars and galaxies. Since galaxies are essentially collections of stars, one expects galaxies to be considerably bigger and brighter than individual stars. A typical galaxy is about 100 billion times brighter than an individual star. It also appears as an extended object in a telescope.

It is, therefore, surprising that there exists an exceptional class of objects, now called 'quasars', which are almost as luminous as an entire galaxy but very compact in size. The discovery of such objects in the early 1960s revolutionized the study of extragalactic astronomy. Nearly a quarter of a century later, today, astronomers are still debating about the exact nature of these objects.

The first quasars were discovered through the radio survey of the heavens carried out in the late 1950s. During these surveys, a radio astronomer, Thomas Mathews, obtained the precise position of a source called 3C 48. (The letters 3C stand for third Cambridge Catalogue.) He noticed that this object seemed to look like a star but its optical spectrum was very peculiar. There were numerous lines in the spectrum, none of which matched with spectra of known elements. A couple of years later, astronomers obtained the exact position of another radio source called 3C 273. Once again, the source had a star-like appearance but a strange spectrum. The lines in this spectrum could be identified with those of hydrogen, provided the redshift was taken to be about 0.158. This means that the wavelength of the radiation which is observed by us is 1.158 times longer than the original wavelength of the hydrogen lines. The spectrum of 3C 48 was soon re-examined, and it was found that all the lines could be identified if the redshift was 0.367.

The redshifts associated with these two radio sources are enormous by normal standards of stellar astronomy. We saw in section 2.5 that the fractional shift in the wavelength is v/c where v is the speed of the source and c is the speed of light, which is about 300 000 km s^{-1}. Stars usually move with speeds of about a few hundred kilometers per second. For a star with a speed of 300 km s^{-1}, the fractional shift will be 0.001, that is, 0.1 per cent. Clearly, the fractional shift of 0.158 of the source 3C 273 would imply that the source must be moving at about 15.8 per cent of the speed of light, which is about 47400 km s^{-1}. Such speeds are unknown for any star or even for galaxies in a cluster. For the source 3C 48 we will obtain still larger values. (In fact, the reason one could not originally identify the lines in the spectrum of 3C 48 was that these lines were redshifted by an enormous amount, which astronomers had not come across before. One of the spectral lines of hydrogen usually found in the visible band had moved into the infrared and hence was not visible; another line usually found in the yellow-green part of the spectrum had moved into the deep red, and so on.)

The most straightforward interpretation of the enormous redshift is based on cosmology. If we assume that the quasars are very far away, then the expansion of the universe will provide such a high redshift. Over decades, people have attempted alternative interpretations for quasar redshift, but no other model seems to work as well as the cosmological model for quasar redshifts. Such high redshifts and large distances necessarily mean that the radiation has taken a long time to reach us. In other words, quasars are the oldest objects known to us and must have formed when the universe was considerably younger. Any model for structure formation in the universe has to account for the observed abundance of quasars when the universe was still fairly young and smooth. It turns out that this fact can put severe constraints on the models for structure formation as well.

The enormous distance attributed to the quasars (based on the redshift) raises an important question. If this object is so far away and is still detectable by us, then it must be extremely luminous. In fact, quasars have to be 100 times more luminous than the brightest galaxies known. By the same token, one would have then expected quasars to be also significantly bigger than the galaxies. However, observations related to their size indicate that the situation is exactly the opposite.

By studying the time variations in the intensity of the emitted radiation, it is possible to determine the size of the region from which the radiation is pouring out. Such studies suggest that some of the quasars must be emitting the radiation from regions which are of the order of 1 parsec or so. This size is nearly 10 000 times smaller than that of a typical galaxy. In other words, quasars manage to emit a tremendous amount of radiation from a very compact region.

The above facts have made theoretical modelling of quasars a difficult task. Conventional sources of energy cannot meet these requirements, and one has to look for more esoteric schemes. Many astronomers believe that quasars are powered by a compact 'central engine' containing a supermassive black hole surrounded by a disc of gaseous material. As the material spirals and falls inside the black hole, viscous processes heat up the matter and can lead to the emission of significant quantities of energy. The conventional view considers quasars as extreme examples of galaxies which some such form of abnormal activity going on in their centres. Such 'active galactic nuclei' have been studied extensively in the last twenty years or so. Several different kinds of esoteric object have been discovered during this period, filling the gap between normal galaxies at one end and quasars at the other extreme end. In spite of these strides in observational astronomy, no-one has yet found a simple explanation for these objects.

7.3   Radio galaxies and active galactic nuclei

Once astronomers realized that galaxies are star systems situated far away from us, they tried to gather as much information as possible about the properties of these objects. Pictures of galaxies taken by optical telescopes showed a variety of shapes (which we discussed in chapter 4, section 6) but did not suggest anything more exotic about these objects. But with the development of radio telescopes — which are capable of detecting radio waves emitted by these objects — the astronomers came in for a real surprise.

It turned out that several galaxies which are fairly uninteresting in the optical band showed remarkable structures in the radio band. For example, there is an object called Centaurus A which is fairly innocuous in the optical band but shows a large jet-like stream flowing out of the galaxy when observed in radio. Several such objects are now known and are called 'radio galaxies'. A typical radio galaxy has an extended structure comprising a central region, two filamentary jets proceeding from the central region in opposite directions and two large lobes at the ends of the jets emitting prominently in the radio band. (There are, of course, several variations in this basic theme.) It is generally believed that the central galaxies are the sites of extremely violent events leading to the ejection of plasma in a jet-like stream. The study of radio galaxies revolves around determining the characteristics of the lobes and their relationship to the central engine, and understanding the physical phenomena which lead to their formation. Though most astronomers believe that we now have a basic understanding of the physics involved, it is only fair to say that the complete picture has not yet emerged.

One dominating feature of such radio galaxies is the large volume occupied by the gaseous plasma, which was presumably ejected by the central galaxy. In fact, the radio lobes are the largest single entities visible to us in the entire universe. In one of the radio sources, called 3C 236, the volume of the radio lobes is nearly 6000 times the volume of the central galaxy. The extent of this object is also large: it is about five megaparsecs in size, which is more than five times the distance between us and the nearest galaxy. This, of course, is an extreme example; but even a run–of–the–mill extended radio source is several hundred kiloparsecs in length. For comparison, the average radius of a galaxy is only about 20 kiloparsec.

What produces the radio emission in these galaxies? Observations suggest that the mechanism is something called 'synchrotron radiation'. Such radiation arises when charged particles move in a magnetic field with speeds close to that of light. The magnetic fields accelerate the particles, making them emit radio waves. From theoretical considerations one can determine the properties of such a radiation, like the shape of the spectrum (which gives the amount of power at different wavelengths within the radio band) and polarization (which measures the direction of vibration of the electromagnetic wave). Many of the radio galaxies show a radiation pattern which is consistent with the theoretical expectation.

The energy radiated in the radio region is around 10^{36} watts in these objects, which is comparable to the energy radiated in the optical region by normal galaxies. In this sense, the radio galaxies are not exceptional. However, in order to emit this kind of energy in the radio band, one needs a vast storehouse of energy in the radio lobes in the form of relativistic particles and magnetic fields. This energy is typically about 10^{60} ergs – which is equivalent to the energy obtained if one million solar masses of material was directly converted into radiation. Based on the extent of the radio lobes, one can also calculate the typical timescale over which these objects should have formed. This age turns out to be between 1 and 100 million years. So the nucleus of the galaxy has to supply an equivalent of 10^{60} ergs in, say, about 10 million years. This requires the production of more than 10^{38} watts of power during the formation stage, which is about 100 times the energy emitted in the optical region of the whole galaxy. It is not easy to explain such violent events, and this remains one of the challenging problems in the physics of radio galaxies.

The timescale associated with the particles which are emitting the radio waves raises another interesting dilemma. We mentioned earlier that these radio waves are emitted by charged particles moving in the magnetic field. As the particle radiates, it will lose energy and soon will cease to be an efficient source of radio waves. In some of the radio galaxies, the timescale in which the energy of the particle will run out is too short. It is, therefore, necessary to think of mechanisms which will re-accelerate the electrons to high energies when they lose a significant fraction of their energy due to radiation. So the problem is not only in producing the necessary amount of energy, but also in supplying it at an appropriate rate to keep the show going.

Radio galaxies are only one kind of a wider class of galaxies known as 'active galaxies'. These are galaxies in the centres of which some kind of violent activity takes place. It is difficult to give a clear-cut definition for an active galaxy, though astronomers have identified several characteristic features. Most active galaxies have a star-like appearance on a short-exposure photograph; they contain strong emission lines in their spectra which will be quite different from that of a normal galaxy; the emission from the central region will show rapid variability, etc. One class of such galaxies is called 'BL Lac objects'. (The name arises from the fact that when the first object of this kind was discovered it was (mistakenly) thought to be a star and was designated as BL Lacertae, that is, variable star BL in the constellation Lacerta, the lizard. Now we know that it is a galaxy and not a star but the name has stuck.) Nearly fifty BL Lac objects are now known, and all of them show violent phenomena in their centres. Another set of active galaxies is known as 'Seyfert galaxies', which were first discovered in 1943, well before the discovery of radio galaxies and quasars. These Seyfert galaxies show strong emission lines coming from their centres, and studies suggest that nearly one per cent of all bright spirals are of this type. They are also strong emitters of infrared radiation, with most of them emitting more than ten times as much in the infrared than our own Milky Way. They also show strong variations in their luminosity and violent explosive phenomena.

All these exotica – quasars, radio galaxies, Seyferts, BL Lac, etc. – share a common feature. In each of them there exists a compact, powerful, energy source in the nucleus which is responsible directly or indirectly for the unusual phenomena. What is happening at the centres of such objects? What is the power house which produces the energy to throw out jets of matter?

To appreciate the enormous amount of energy involved, notice that the power output of a quasar is equal to that of a million supernova explosions per year. Considering the compactness of the region, we will need to pack a billion such stars inside the size of the solar system to produce such energy outputs. Of course, it is unrealistic to think of supernova explosions as powering a quasar. But we can take a clue from the above discussion. If we pack an enormous amount of mass, probably 10^9 $M_\odot$, into a compact region, it is likely to collapse and form a black hole. Astronomers believe that the centres of active galactic nuclei host very compact but supermassive black holes.

A black hole, by itself, cannot radiate too much energy. The actual scenario is more complicated. The black hole exerts an enormous gravitational force on the surrounding gas and stars and makes matter accrete on to it. The accreting gas will not simply fall down the black hole but will orbit around it, gradually spiraling in as it loses energy by radiation. This will lead to the formation of an 'accretion disc' around the black hole, which is thin near the centres and thicker near the outer regions. Such an accretion disc can be a strong source of radiation, especially X-rays. The enormous flux of X-rays will interact with charged particles in the plasma and will accelerate them. The energetic particles will find it difficult to escape through the thick disc, and hence will be emitted perpendicular to the plane of the disc. This will produce two jets of high speed particles flying away from the nucleus. As they penetrate the interstellar medium of the galaxy, they become narrow and collimated. One will expect emission of X-rays and high energy ultraviolet radiation from the disc as well as a jet-like stream. Most of the active galactic nuclei show evidence for such emission (see figure 7.2).

The region surrounding the central power house will contain a fair number of fragmented gas clouds which are orbiting the nucleus at speeds of several thousand kilometers per second. When high energy X-rays hit these clouds, they will be heated up and will radiate normal emission lines, especially hydrogen lines. But since the clouds are moving with random velocities, amounting

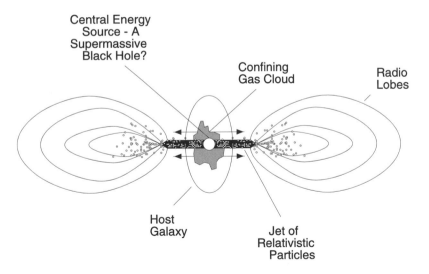

Figure 7.2   The current model for some of the radio galaxies is based on jets of particles being emitted by the nucleus of an active galaxy. These particles are eventually stopped by the gas in the intergalactic medium and in the process produce large radio lobes.

to a few thousand kilometers per second, the lines will be much broader than what is usually seen. Such a region around the nucleus is called a 'broad–line' region, and needs to be only about one parsec or so in size. Still further away, at about one kiloparsec or so, there will be low density clouds orbiting with smaller speeds. They will be emitting narrower lines of hydrogen. In fact, models like this have been suggested to explain in a unified manner the phenomena of radio galaxies, quasars, etc. In such models, the different objects arise from the same central source, depending on the angle at which we are viewing the source.

While quasars and active galaxies are the farthest objects we have seen, it is also of interest to ask: what is the distance to the closest quasar? Alternatively, we may enquire whether quasar formation was a phase in the evolution of the universe or is still continuing. To answer such questions, one needs to make a systematic survey of all quasars in the sky and try to see how the quasar population changes with redshift. Such studies indicate a very interesting picture. It seems that the number of quasars rapidly declines at redshifts lower than about 1.5, and the quasar activity seems to have peaked around a redshift of probably 2.5. There is also an indication that quasar activity has declined at higher redshifts, though this is somewhat more difficult to establish conclusively with current observations. If we take these observations at face value, it follows that our universe went through a 'quasar phase' during $z = 3$ to $z = 1$ (see figure 7.3).

Why is it that quasar activity declined in the universe both at high redshifts and at low redshifts? If the conventional picture of quasars is correct, then quasar activity could have only taken place after galaxy formation had progressed to a certain extent, massive black holes had been formed in the centres of galaxies, etc. This could possibly account for the fact that there were fewer quasars at higher redshifts. It is not very clear why quasar activity declined at lower redshifts, say at $z < 1$. Cosmologists have come up with different explanations for this phenomenon but the issue is still unsettled.

7.4 Quasar spectra and the intergalactic medium

Since quasars are the farthest objects seen in the sky, they can be used as an interesting kind of cosmological probe. The light which we receive from quasars passes through layers of cosmic material in the universe and hence can provide us with information about this material.

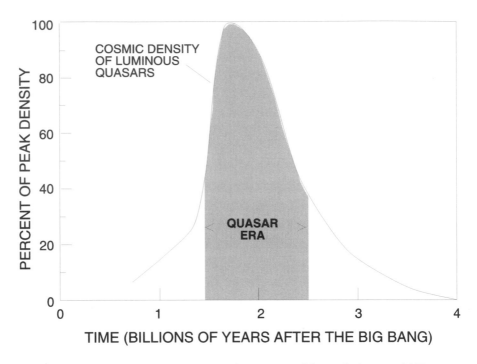

Figure 7.3   Observations suggest that quasar activity peaked at a redshift of about $z = 2.5$ and declined at both higher and lower redshift. This suggests that our universe went through a 'quasar era' sometime in the near past.

In chapter 2 we discussed the emission and absorption of radiation by a hydrogen atom in terms of the jumping of the electron between different energy levels. Consider, for example, the radiation emitted when the electron jumps from levels 2, 3, . . . to the ground level, which is level 1. Each of these will produce a line at a characteristic wavelength in the spectrum of hydrogen, and these are called the 'Lyman series' of spectral lines. The first among them is called the Lyman-α (pronounced 'Lyman alpha') line and occurs at 1216 Å, which is in the ultraviolet. Because of the ozone layer in the atmosphere, this line cannot be seen in, say, the Sun's spectrum when observed from the ground. It was first detected in the Sun in 1959, by the UV detectors which were taken above the stratosphere in rockets.

Because hydrogen is the most abundant element in the universe and the Lyman-α is a prominent line in hydrogen spectra, one would expect quasars to be strong emitters of Lyman-α radiation. What is more, if the redshift of a quasar is large enough, then the Lyman-α line (which is in the UV) will move into the visible part of the spectrum, and could be studied from the ground. It turns out that the minimum redshift needed for such a shift to occur is about 1.7.

This expectation is indeed realized; the spectra of high redshift quasars do show a prominent Lyman-α emission line, which stands out loud and clear. More interestingly, there are several weaker, dark, absorption lines at shorter wavelengths (on the blue side) of this line. These lines arise due to the absorption of the quasar radiation by gas clouds located between us and the quasar. These gas clouds are at different distances from us and hence have different redshifts; this causes the quasar radiation to be absorbed at different wavelengths (see figure 7.4).

Such absorption spectra of quasars allow us to probe systematically the contents of the universe between the quasar and us, and they have been studied

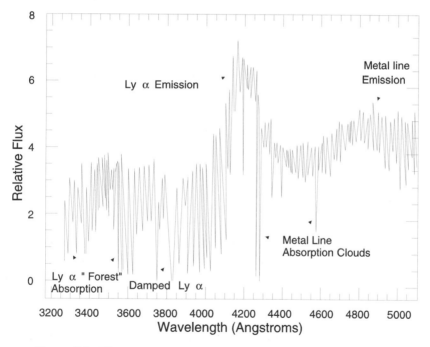

Figure 7.4 The spectrum of a quasar at a redshift of $z = 2.43$ shows several absorption features. The peak in the middle is due to Lyman-alpha emission, which – in the absence of any redshift – occurs at the wavelength of 1216 Å. Here we see it occurring at a redshifted wavelength of $1216 \times (1 + z) = 4170$ Å. The host of absorption lines seen on the left side of this peak are thought to be due to clouds of neutral hydrogen intervening between the quasar and us.

extensively in the last two decades. The Lyman-α absorption lines themselves can be divided broadly into two separate categories. The first category is made of a large number of narrow absorption lines which are seen so densely in the spectra that they are called a 'Lyman-α forest'. The second category is made of broad, deep absorption troughs called 'damped Lyman-α lines'.

The study of Lyman-α forest lines allow us to understand the physical conditions of these clouds, like the abundance and density. The size of the clouds can be estimated from a different procedure. In some cases, very similar patterns of Lyman-α forest lines are seen in the spectra of two quasars which are close to each other in the sky. This is most naturally explained if some of the intervening clouds are big enough to cover both the quasars. Such estimates suggest that a typical Lyman-α forest cloud could be as large as a few hundred kiloparsecs. These clouds also show different properties at different redshifts, suggesting that they participate in cosmic evolution. Some of these clouds contain no heavy elements and are very weak in absorbing the light. This implies that the amount of neutral hydrogen in these clouds is very low, and in fact lower than in any of the interstellar clouds in our galaxy.

There are other clouds which show absorption due to elements heavier than hydrogen. In fact, mounting evidence suggests that a significant fraction of the clouds which are in the line of sight of a quasar contain heavy elements. Such an observation is extremely important, since these heavy elements could have been manufactured only by the first generation of stars. So if we see traces of heavy elements at high redshifts, then we must conclude that some amount of star formation should have taken place at epochs earlier than that. This offers a vital clue in deciding when the first generation of stars formed in our universe. These metal-rich absorbing clouds are probably interstellar clouds of distant galaxies which have come into the line of sight. The galaxies themselves could be so distant and faint that they are not seen against the bright glare of the quasar. In some cases, very careful imaging of the region has revealed faint traces of the intervening galaxy, lending support to this view.

The second category of clouds, called 'damped Lyman-α' clouds, are very similar in structure to a gas-rich disc galaxy viewed at a random angle. For example, the Milky Way galaxy in its formative stages could have looked like one of these clouds. The absorption spectra due to these clouds also show the presence of heavy elements, but with an abundance which is only about 1 to 10 per cent of that in the Sun. This agrees with the conjecture that these damped Lyman-α systems are precursors of galaxies and formed early in their evolution.

The study of quasar spectra also provides us with another vital piece of information about the material which lurks between the galaxies, called 'the intergalactic medium'. The models discussed in the last chapter suggest that structures in the universe formed out of the collapse of primordial gas under self-gravity. However, such a process can never be one hundred per cent efficient. A large fraction, say ninety per cent, of primordial hydrogen and helium could have collapsed to form galaxies, but ten per cent could be left behind as a smoothly distributed gas. If this is true, the quasar light reaching us has to swim through this smooth gas of neutral hydrogen. All the radiation at sufficiently high frequencies will be absorbed by such a smooth distribution of gas, which is spread everywhere from the quasar to us. This would lead to a clear signature in the spectra of quasars, using which one could estimate the amount of neutral hydrogen in the intergalactic medium.

Observations, however, lead to a very surprising result. No signature for absorption by *smoothly* distributed hydrogen in the intergalactic medium was seen. More precisely, the study of quasar spectra showed that the amount of smoothly distributed neutral hydrogen at a redshift of $z = 3$, say, is nearly a million times lower than the mean density of nucleons in the universe at that epoch. Such a result allows only two possible interpretations. It may be that galaxy formation is so extraordinarily efficient that, by $z = 3$, virtually all matter has formed compact self-gravitating units, and no neutral hydrogen is distributed smoothly in the intergalactic medium. This interpretation, however, does not stand up against further scrutiny. No model for structure formation could be so efficient. The second alternative is that all the matter which is present in the intergalactic medium is not neutral but ionized. The observations based on absorption spectra, of course, could only detect neutral hydrogen; if the hydrogen is ionized in the form of free protons and electrons, the observations will not put any constraints on that. Most cosmologists believe that this is indeed the case, and that the intergalactic medium is ionized.

Such a picture of the intergalactic medium leads to several far-reaching conclusions. To begin with, you will recall that neutral atoms were formed in the universe at a redshift of $z = 1000$. If we accept that the intergalactic medium at low redshifts had no neutral atoms but only ionized plasma, then this ionization should have taken place at some time between $z = 1000$ and, say, $z = 5$. (The second bound, $z = 5$, comes from the fact that we have quasar redshifts ranging all the way up to about $z = 5$ and hence we can probe the intergalactic medium up to this redshift.) What physical process could have ionized matter between these redshifts? And, when did this process occur?

There are no clear-cut answers to these questions at present. Most astronomers believe that at a redshift of, say, $z = 8$ or so, the first structures must have formed in our universe. These precursors to our present-day galaxies must have gone through a phase of massive star formation, and the UV radiation from these first generation stars must have ionized the intergalactic medium. In order for this scenario to work, it is necessary for some kind of structures to form at sufficiently high redshifts. From the observations it is clear that this should have occured at redshifts greater than 5. As we shall see in the next section, this is not an easy matter to arrange.

Secondly, the above observations suggest a rather complex structure for the intergalactic medium. We had already seen that there are clouds of neutral hydrogen gas between us and the quasars. These clouds have to be necessarily embedded in a much hotter plasma of ionized gas. The dynamics of such a two-phase medium can be quite intricate. For example, it is not very clear what holds the Lyman-α clouds together. One possibility is the gravitational force due to a dark matter halo around the cloud. Another possibility is the pressure due to the ionized medium which exists outside the cloud. In fact, we do not even know for sure whether the clouds are nearly spherical, isolated blobs or much more extended sheet-like structures spread through the universe. Computer simulations are just beginning to address such questions, and most cosmologists believe that a more definitive picture will emerge in the next few years.

In the above discussion, we concentrated on the intergalactic medium which exists in the background universe. There is another kind of intergalactic medium about which astronomers have gathered a considerable amount of data in the last few years. This is the medium which exists between galaxies *in a cluster* and is usually called 'intracluster gas'. We saw in section 4.8 that the intracluster gas is very hot, with a temperature of about 100 million degrees kelvin. Such a hot gas emits a copious amount of X-rays, the study of which allows us to probe the amount of gas. It turns out that nearly ten per cent of the mass of clusters is in the form of intracluster gas, and about ten per cent is in the form of visible galaxies; nearly eighty per cent of matter in clusters is dark.

This is not the only surprise which the intracluster medium has in store. It turns out that, even at 100 million degrees, the iron atom is not completely ionized but retains one or two of its electrons. The transitions of these electrons produce spectral lines in the X-ray band. By studying these spectral lines, one can estimate the amount of iron in the intracluster medium. This abundance of iron, relative to hydrogen, turns out to be nearly one third of that in the Sun.

Such a high abundance can only arise if the intracluster gas has been processed by stars and subsequently ejected by the galaxies. The structure formation history of galaxies and clusters should be able to explain these features in order to be considered satisfactory. This turns out to be a somewhat difficult task.

7.5   High redshift objects and structure formation

We saw in the last chapter that structures bound by self-gravity can form in large numbers only when the density fluctuation at the relevant scale reaches a value of about unity. If we know the masses of different astrophysical objects, then we can compute roughly the epoch at which such objects would have formed in any given model for structure formation. For example, the amplitude for density fluctuation at galactic scales (with a mass of of 10^{12} $M_{\odot}$) reaches a value of unity around $z \approx 3$ in the pure CDM model. So in that model copious production of galaxies can take place at a redshift of $z < 3$. Structures with larger mass will form later and structures with less mass would have formed earlier.

There is, however, one important caveat to the above statement. To understand this, consider a redshift of, say, $z = 6$. Most objects which are forming in the universe will have a mass of about 10^{10} $M_{\odot}$, i.e., they will be subgalactic. This, however, does not mean that we will not be able to form even a single galaxy-like object at this redshift. A more precise calculation will show that a tiny fraction of the objects which are formed can have masses as high as 10^{13} $M_{\odot}$, even at this redshift.

The importance of this analysis to observations related to the high redshift universe is the following: if we find a large number of galaxy-like objects at a redshift of 6, then we may have to seriously question the validity of CDM models. But if we see a few, rare, galaxy-like objects at a redshift of 6, it may be possible to accommodate them in the CDM model. What is important is the abundance of structures, viz. how many objects are there per unit volume, at a given redshift. Given a model for structure formation, we can compute what fraction of the structures which are forming at a given redshift will have a certain value for their mass. At higher redshifts the most abundant structures will have lower masses. If we can observationally determine the abundance of different structures at high redshifts, then we can compare it with the theoretical results, thereby putting fresh constraints on the models.

It turns out that such an analysis spells doom for several models for structure formation which are otherwise quite successful. As an example, consider the two models described in chapter 6, section 6.9, the mixed dark matter (MDM) model and the one with the cosmological constant. Both these models were *designed* to have smaller density fluctuations compared to pure CDM models at galactic and subgalactic scales (see figure 6.7). This means that the density fluctuation at the scale of dwarf galaxy or galaxy will reach the value of unity at a later time in these models compared to pure CDM models. If galaxies form copiously at $z = 3$ in CDM models, they can form copiously only at, say, $z = 1$ in the MDM model. In short, structure formation is delayed in these two models compared to the CDM model for the simple reason that they have less density contrast than the CDM model at the relevant scales.

We can state the same result in terms of the abundance of different structures. Galaxy-like structures will be quite abundant at $z = 3$ in the CDM model but will be comparatively rare in the other two models. In fact the situation turns out to be a bit of a disaster. The MDM model cannot correctly account for the observed abundance of damped Lyman-α absorbers seen in the spectra of quasars. Hence, it is very likely that this model is not viable. The model with the cosmological constant can just barely accommodate the observed abundance of damped Lyman-α systems. Because of the uncertainties in the observations and theoretical modelling, one cannot yet rule out this model, but its existence is severely threatened. With improved observations one will be in a better position to make a more definitive statement. It will also be possible, eventually, to use the abundance of quasars and radio galaxies to put similar constraints; at present, the observations are not good enough.

The situation outlined above is actually quite generic. Given the COBE observations, one can fix the amplitude of density fluctuations at very large scales. In order to obtain the correct correlation function for galaxies, we have to lower the amplitude at small scales compared to pure CDM models. But if we do that we will necessarily delay the formation of structures. Note that these two observations correspond to vastly different length scales: the COBE observations probe length scales of thousands of megaparsecs, while the galaxy correlation function probes 16 Mpc or so. They also describe the universe at vastly different redshifts: COBE probes the condition at $z = 1000$ while galaxies essentially sample the universe at $z = 0$. Models which satisfy conditions at these extremes may or may not succeed in describing other features. For example, the abundance of damped Lyman-α systems probes the universe at the redshift range of $z = 2 - 4$ and is sensitive to density fluctuations at subgalactic scales. Of the two models discussed above, the MDM model fails the test, while the model with the cosmological constant just barely passes. It is remarkable how the host of observations which are currently available can put a strangle-hold on theoretical models.

7.6 Gravitational lensing

In Einstein's theory of gravity, all forms of energy are influenced by the gravitational field, since matter and energy are interconvertible. A beam of light travelling close to a massive gravitating object, for example, will be pulled towards it. In fact, such a 'bending of light', when starlight was passing close to the Sun, was one of the crucial tests of Einstein's general relativity. This effect can have a more dramatic and useful manifestation in the form of 'gravitational lensing'. Gravity can act as a powerful telescope, allowing us to probe objects which would have been otherwise inaccessible.

The simplest version of such a lensing action is shown in figure 7.5. Light from a distant quasar is influenced by the gravitational field of, say, a galaxy before reaching the Earth. Because of the gravitational pull of the intervening galaxy, the light beam is bent as shown in the figure. Under favourable circumstances, two separate beams of light from the quasar can reach us, travelling along two separate paths as shown. This will give rise to two images at A and B. The actual path followed by the light from the quasar to Earth is marked as 'path 1' and 'path 2'. We will, however, think of the light as coming along the lines AC or BC, giving rise to the two images.

Since A and B are images of the same quasar, they will both show identical characteristics like spectra, etc. But since the paths 1 and 2 taken by the light beam are different, the light will take slightly different times to reach us along these two paths. This means that any time variation in the intensity of the quasar will not appear in the images 1 and 2 simultaneously but will exhibit a time delay. Further, just as with any other image formed by a lens, the image can be a magnified version of the original quasar. This opens up the possibility of us being able to observe faint objects through the magnifying action of a gravitational lens.

Astronomers have detected several examples of such gravitational lenses in the universe. In each case, the detailed study of the images allows one to understand several features of the intervening mass distribution (like the galaxy) which acts as the lens. In fact, the description given above is only

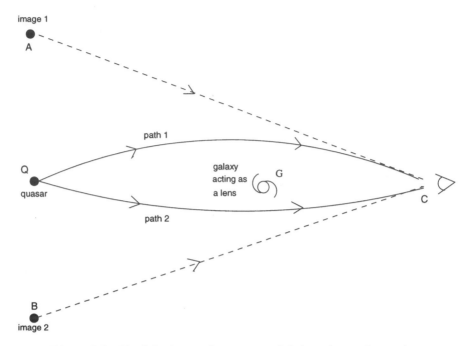

Figure 7.5 The light from a distant quasar Q is bent due to the gravitational field of an intervening galaxy G and reaches us at C. Under favourable circumstances this can happen along path 1 or path 2, leading to two images of the quasar Q in the directions CA and CB.

the simplest scenario which is possible. In general, gravitational lensing can lead to not just two images but many more. One dramatic example is a quasar at a redshift of 2.55 called H 1413+117 which leads to a system of 4 images in the shape of a 'four-leaf clover' (see figure 7.6). The images are separated from each other by about 1 arcsecond in the sky. Modelling such a complex lens system is a difficult task, and any successful model will tell us a lot about the mass distribution in the universe. In general, galaxies and clusters of galaxies act as gravitational lenses, and sometimes the lensing action is due to a combination of more than one intervening galaxy.

Gravitational lensing can also manifest itself in a different manner. You must have noticed that objects viewed through a lens appear distorted. In the same way, the gravitational field of the intervening matter distorts the image of the distant objects. When the intervening matter is sufficiently dense this leads to the occurrence of multiple images of the source, as discussed above. But if the intervening matter is somewhat less dense, one still sees some effect. Instead of

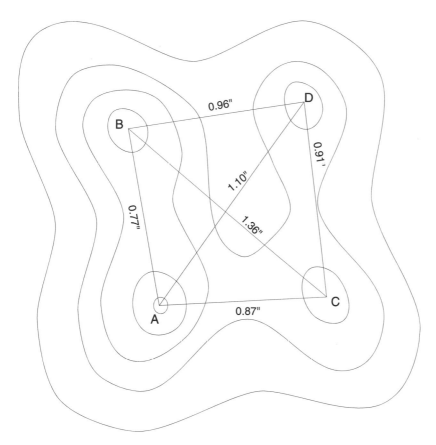

Figure 7.6   In general, gravitational lensing can lead to multiple images with complex geometry. In the case of the quasar H 1413+117, lensing leads to four images separated from each other by about 1 arcsecond.

multiple images, we get to see large, arc-like, images of distant galaxies. This phenomenon, called 'weak lensing' is usually caused by an intervening cluster of galaxies acting as a gravitational lens.

Astronomers have seen evidence for such arc-like objects in several cases. These arcs are quite large and can stretch to distances of about a few mega-parsecs in length and 100 kpc or so in width. The first set of almost perfect arcs was discovered in the mid-1980s and – after some initial debate – was identified as a result of gravitational lensing by an intervening cluster of galaxies. One of the best examples of such arcs lies in a cluster of galaxies known as Abell 370. This cluster is at a redshift of 0.4 or so, while the light from the arc has a redshift of about 0.7. In other words, the light from the arc actually originates from a distance which is twice as large as the distance to the cluster. This arc-let is considered to be a highly distorted image of a far-away galaxy.

Theoretical modelling of the lensing cluster in such cases invariably leads to the conclusion that they contain vast quantities of dark matter. The luminous matter in the cluster is woefully inadequate to produce such lensing effects. In fact, it is possible to actually obtain information about the distribution of gravitating matter from the distorted images of distant galaxies. By comparing the distribution of gravitating matter (which is dominated by dark matter) with visible matter and X-ray emitting gas in the cluster, one can obtain valuable information about the distribution of visible matter vis-à-vis the dark matter.

Recently, gravitational lensing has found yet another use in cosmology. Astronomers have always conjectured the possible existence of compact but massive objects in the halo of our galaxy. These could, for example, be dead relics of stars, black holes or even planet-like objects. Since these objects are invisible, they should also be properly considered as dark matter, and it is important to know whether they exist and if so how much mass they contribute to the halo. It turns out that such Massive Compact Halo Objects (MACHOs, for short) can be detected through their gravitational lensing action. The idea is the following: when a MACHO moves across the line of sight to a distant star, the gravitational lensing action will temporarily brighten the background star. By measuring this brightening effect, one can figure out the properties of MACHOs. There are, however, two difficulties. Firstly such an event in which a MACHO comes across the line of sight to a star (called a 'microlensing event') is very rare; the chances are roughly 1 in 2 million. Secondly, many stars have intrinsic variation in their brightness, which has to be distinguished from the brightening due to microlensing. The first difficulty can be overcome by monitoring several million stars in the Large Magellanic Cloud (which is a satellite galaxy to the Milky Way; see section 4.8). The second difficulty can be circumvented by looking for some very specific signatures of microlensing and careful analysis of data. Such observations have been in progress for a few years now, and several such microlensing events have been detected. The preliminary analysis of data shows that nearly 10 per cent of the dark matter in the halo could be in the form of MACHOs. Over a period of time this analysis will improve and we will be in a better position to understand how much of dark matter is made of MACHOs and how much by WIMPs.

Finally, gravitational lensing also provides an interesting constraint on one of the models for structure formation discussed earlier which has survived all the observational tests, viz. the one containing CDM and a cosmological constant.

One direct test of such a model would be verification of the existence of a non-zero cosmological constant in our universe. It turns out that the angular separation between the lensed images of quasars depends on the parameters of the background cosmology; so the separation between the images will be different in a universe with a cosmological constant compared to another with zero cosmological constant. More precisely, each model will provide us with a distribution of possible angular separations between the images. Comparing this distribution with the observed separation of lensed images, one can try to discriminate between different models. In principle, such a procedure could allow us to determine the value of the cosmological constant. In practice, observational errors prevents the successful completion of such a program. But even with available observations (and their uncertainties) one can claim that the cosmological constant in the universe cannot contribute more than about 70 per cent of the critical density. The model for structure formation based on the cosmological constant requires roughly this value in order to satisfy all observations. So we see that, at present, this model is living rather dangerously. With improved observations, the verdict could go either way.

8

Open questions

8.1   The broad picture

In the previous chapters, we have explored the conventional thinking of cosmologists and astrophysicists in their attempt to understand the structures in the universe. Some of these attempts have been very successful, while others must be still thought of as theoretical speculations. Since different aspects of structure formation were touched upon in different chapters of this book, it is worthwhile to summarize the conventional picture in a coherent manner.

The key idea behind the models for structure formation lies in treating the formation of small-scale structures like galaxies, clusters, etc. differently from the overall dynamics of the smooth background universe. This is linked to the assumption that, in the past, the universe was very homogeneous with small density fluctuations.

The evolution of the smooth universe is well described by the standard big bang model. Starting from the time when the universe was about one second old, one can follow its evolution till the time when matter and radiation decoupled – which occured when the universe was nearly 400 000 years old. During this epoch, the energies involved in the physical processes ranged from a few million electron volts to a few electron volts. This band of energies has been explored very thoroughly in the laboratory experiments dealing with nuclear physics, atomic physics and condensed matter physics. We understand the physical processes operating at these energy ranges quite well, and it is very unlikely that theoretical models based on this understanding could go wrong. In other words, we can have a reasonable amount of confidence in our description of the universe when it evolved from an age of one second to an age of 400 000 years.

This big bang model – which was described in chapter 5 of this book – also makes two major predictions about the universe. Firstly, it describes a phase of primordial nucleosynthesis in which helium and deuterium were synthesized in the universe. We saw in chapter 5, section 2 that about 24 per cent of mass in the universe would be in the form of helium nuclei when it was more than a few minutes old. The amount of deuterium which was synthesized is a sensitive function of the nucleonic density and suggests that the nucleonic density in the universe is less than 10^{-30} gm cm^{-3}. Secondly, the big bang model describes an evolving universe in which matter and radiation were in thermal equilibrium during the early epochs. A remnant of this era would be a characteristic thermal radiation surrounding us today. Such a radiation, called the microwave background radiation, was indeed discovered in 1965, and it provides the strongest support for the big bang model available today.

Given that our model for the smooth background universe is basically correct, we have to address next the question of how various structures have formed. In such a study, we are aided by direct observational evidence as regards the formation of stars. While theories for star formation (or planet formation, for that matter) are not yet completely free of controversies, most astronomers think that the outstanding issues in these areas are only matters of detail. For this reason, we have concentrated mostly on the question of *galaxy* formation in this book. To understand the formation of galaxies and larger structures, one invokes the concept of gravitational instability. This paradigm is based on the fact that, under the action of an attractive gravitational force, any density fluctuation would be enhanced, and structures could eventually form out of such growing density perturbations. If the early phases of the universe contained small perturbations in density, the growth of such density perturbations can – in principle – lead to the kind of structures we see today.

Once again, most cosmologists believe that this idea is basically correct. In fact, it is hard to come up with any other viable mechanism for the growth of structures in an expanding universe. What is more, this idea makes a strong prediction: the cosmic microwave background radiation must contain small temperature fluctuations which are the direct result of density fluctuations that existed when the universe was nearly one thousand times smaller. This prediction was made in the late 1970s, and a detailed calculation of the temperature anisotropy based on different models for structure formation was available in the 1980s. Such temperature fluctuations were detected by COBE in 1992, and the theoretical models received strong observational support, as far as the overall picture is concerned.

To form the complete picture, one also needs to know the material content of our universe. Direct observation in different bands of the electromagnetic spectrum gives us information about the visible universe made of normal matter (protons, neutrons and electrons) and radiation (photons). But, as we saw in chapter 5, we have indirect but very firm evidence for the existence of some form of dark matter in the universe. In fact, these observations suggest that there is nearly 10 to 100 times more dark matter than visible matter in the universe. It is, therefore, necessary to postulate some of the key attributes of the dark matter while developing models for structure formation. It would have been wonderful if we had direct laboratory evidence for the dark matter particle. May be we will have it within the next decade; but, in the absence of such evidence, one needs to proceed by making reasonable assumptions regarding the nature and distribution of dark matter.

Given the two ingredients described above – a smooth, expanding background universe and the existence of small density perturbations in the past – one can sketch the overall evolution of our universe along the following lines. For the sake of convenience, we can divide the evolution into different stages.

Stage 1: The very early universe

This corresponds to the phase of the universe at which the mean energy of the particles was higher than about 100 GeV. The highest energy which can be explored in the man-made accelerators today corresponds to about 100 GeV. So we have no direct observational evidence for the physical processes which could have taken place in the universe during this stage. There are, of course, several theoretical *speculations* about this stage, especially since this stage of the universe plays a vital role in deciding two key inputs: firstly, the universe emerged out of this stage with a distinct asymmetry between matter and anti-matter; there was an excess of matter over antimatter when the universe reached the end of stage 1. Most cosmologists believe that high energy processes taking place at this stage could be responsible for this asymmetry. Secondly, the universe at the end of stage 1 is most likely to have inherited small density fluctuations, which – eventually – could have led to the formation of structures. It is, of course, possible for such fluctuations to have originated at a later stage, but most popular models arrange the origin of these fluctuations within stage 1. If the fluctuations originated during the inflation of the universe (see chapter 5, section 4) then they must definitely have originated in stage 1.

Stage 2: The era of nucleosynthesis

The second stage covers the period when the mean energy of particles in the universe changed from 100 GeV to about 1 keV. During this epoch, the universe expanded and cooled as a homogeneous entity, with the fluctuations playing a very small role in the dynamics. Two major events took place in this stage. To begin with, most elementary particles disappeared by annihilation with antiparticles. (For example, the universe would have contained a large amount of muons and antimuons at the end of stage 1. When the temperature fell below about 10^{12} K, the photons became incapable of producing muon–antimuon pairs, while the existing pairs annihilated and disappeared.) At the end of stage 2, we are left with just protons, neutrons, electrons, neutrinos and photons among the familiar particles. If the dark matter is made of some exotic weakly interacting massive particles (called 'WIMPs') then WIMPs will also have existed at the end of stage 2.

The second most important event which took place during this stage was the synthesis of atomic nuclei. As the universe cooled below the binding energy of atomic nuclei, it became possible to form the nuclei of deuterium and helium out of the primordial soup, with helium contributing to about 24 per cent by weight. What is probably more important, the expansion of the universe prevented the synthesis of heavier elements in significant quantities. Since nucleosynthesis requires high temperatures, the heavy elements we see today need to have been formed in some other hot cauldrons in the universe. (We know that this synthesis took place in the cores of the stars). At the end of stage 2, the universe was made of helium nuclei, hydrogen nuclei, electrons, neutrinos, photons and WIMPs.

Stage 3: The era of atoms

The third stage spans the period when the temperature of the universe changed from about 10^7 K (which was the temperature at the end of stage 2) to 2500 K. Most of the interesting events in this stage took place towards the end. When the temperature of the universe was about 10^4 K, the energy density of the radiation fell below the energy density of matter and the universe made a transition from being radiation dominated to being matter dominated. Most of the energy density in matter, of course, is contributed by the dark matter particles which we have called WIMPs. Once the universe became matter dominated, the density fluctuations in the WIMPs started growing due to gravitational instability, while the density fluctuations in normal matter (made of nuclei and electrons) remained frozen because normal matter was tied to the radiation. As the universe cooled still further, and reached a temperature of about 2700 K, the stage was set for the formation of neutral atoms. The electrons combined with protons to form hydrogen atoms and with helium nuclei to form helium atoms. Once this recombination took place, photons ceased to interact with matter and flowed freely through space. The matter, in turn, was now influenced by the gravitational force of the WIMPs, and the density fluctuations in normal matter got equalized to that of dark matter very quickly. These density fluctuations, however, were still small and were about 1 part in 10 000.

The physical processes which took place during stage 3 allow us to make two concrete predictions about our universe. Firstly, the photons which decoupled from matter should be reaching us today with a characteristic thermal spectrum. Secondly, this thermal spectrum should contain small temperature fluctuations of about 1 part in 10^5 which should be detectable. The fact that observations have vindicated both the predictions can be taken to be a major success for our basic model.

Stage 4: The unknown era

The period during the history of the universe about which we have no direct knowledge corresponds to the time when the universe cooled from 2500 K to about 16.2 K. In terms of redshifts, this covers the period between the redshifts of about $z = 925$ (when stage 3 ended) and $z = 5$. The microwave radiation allows us to probe the universe at $z \approx 1000$, while we have direct observational evidence of different structures in the universe at redshifts below $z = 5$. But for the period in between, we have to rely heavily on indirect circumstantial evidence. Knowing what happened in stage 3 and from the observations at low redshifts, one has to reconstruct this phase.

One can indeed make some intelligent guesses about the universe during this stage. First of all, the contents of the universe are fairly well determined. It is made of a gaseous mixture of hydrogen and helium interacting with the WIMPs which form the dark matter. Secondly, the theory of gravitational instability predicts that density inhomogeneities must have grown significantly during this epoch. During stage 4, the universe expanded roughly by a factor of

200 or so; based on our discussion in chapter 6 (see section 6.6) we may conclude that the density fluctuation would have also grown by typically the same factor. It is also possible that sub-galactic structures made of about 10^3 to 10^6 $M_\odot$ had already formed during this epoch. This result, however, depends very much on the theoretical model which is used and the pattern of initial fluctuations. But if this is indeed true, then the first generation of luminous objects could have formed around the end of stage 4. They could be very massive stars with a very short lifetime emitting copious amount of radiation. This radiation will have ionized most of the matter in the universe which is populating the region between the galaxies today. Observations of the high redshift universe, discussed in chapter 7, do suggest that the intergalactic medium is ionized; the above scenario could possibly provide the reason for this feature.

Stage 5: The high redshift universe

The universe between redshifts of $z = 5$ and $z = 1$ is covered by this stage. We have direct observations of this high redshift universe, and it is only a question of time before astronomers and cosmologists put together various pieces of the jig-saw puzzle. The universe in this redshift range is dominated by quasars, radio galaxies and other active galactic nuclei. But this impression arises partly from the fact that such objects are highly luminous and could be detected with relative ease by our telescopes. A normal galaxy (like the Milky Way) or a cluster (like Coma) would be hardly detectable even if they had existed at these redshifts. This fact raises the question: when did galaxy-like structures first form? The observations of the high redshift universe suggest that some kind of galaxy formation was *certainly* going on during this stage, if not even earlier.

It is also around this stage that one expects segregation between dark matter and visible matter. Normal matter can form luminous structures, emit radiation and cool, none of which dark matter can do. The cooling of normal matter allows it to fragment and sink to the central regions of the potential well, made of dark matter. Further morphological division of normal matter – like elliptical galaxies and spiral galaxies or as dwarf galaxies and supermassive galaxies – takes place essentially due to complex gas dynamical processes. The difficulty in understanding this phase is not so much at the conceptual level as in making sure that all the necessary physical processes are properly handled in the models for structure formation. Massive computers capable of high speed computation are already on their way into throwing some light on the hydrodynamical processes which took place in this stage. Within the next decade or so we will probably have a much clearer view of the high redshift universe.

Stage 6: The nearby universe

The epoch between redshifts of $z = 1$ and $z = 0$ is dominated by nearby cosmological objects, and in some sense constitutes what we call the 'present day' universe. Most of the structure formation reaches a well defined pattern in this phase, and we have a fairly clear indication of clustering in the form of galaxies, groups, clusters and possibly superclusters. Galaxy surveys have shown that the universe is punctuated by matter distributed along sheet-like and filamentary structures. Observations of dark matter suggest that these visible structures are embedded in a more diffuse halo of dark matter. The theoretical understanding of this phase is again limited only by our inability to handle complex hydrodynamical processes.

Much of the star formation, genesis of planets (not to mention the growth of human civilization) took place during this stage. These processes, again, are not directly influenced by large-scale cosmic phenomena; we see evidence for star formation in different regions of our own galaxy which is entirely governed by the physical processes taking place in those regions. As we emphasized before, the theories dealing with star formation or planet formation are still at their infancy, but it is likely that they are not seriously influenced by the vagaries of structure formation at larger scales. Astrophysics, rather than cosmology, starts to dominate the scene at this stage.

The various stages are summarized in table 8.1. Given this broad picture of structure formation, we can ask the question: what are the key results in which one can have a reasonable amount of faith? And, what are the areas which are still in the domain of theoretical speculations? We now turn to these questions.

8.2   A critique of structure formation

In the conventional approach to structure formation, we divide the universe into a smooth background punctuated by matter clustered at different scales. The first question which springs to one's mind is whether such a modelling is

Table 8.1. *Cosmic calendar*

Stage	Redshift	Temperature	Description
1	$z > 10^{14}$	greater than 10^{15} K	Quantum universe; inflation; generation of density fluctuations.
2	$10^{14} > z > 10^{6}$	$10^{15} - 10^{7}$ K	Synthesis of helium and deuterium; disappearance of heavy elementary particles.
3	$10^{6} > z > 925$	$10^{7} - 2500$ K	Formation of neutral atoms; decoupling of radiation from matter.
4	$925 > z > 5$	$2500 - 16.2$ K	Growth of density fluctuations; possible formation of first structures; ionization of intergalactic medium.
5	$5 > z > 1$	$16.2 - 5.4$ K	Segregation of nucleonic and dark matter; formation of galaxies and clusters; era of quasars.
6	$1 > z > 0$	$5.4 - 2.7$ K	Clustering of structures; copious star formation; genesis of planets and life.

correct or whether one should take a more holistic approach to cosmology. Observations seem to favour the division between a smooth background cosmology and the existence of small-scale structures, which could have grown out of gravitational instability. We have reviewed several of these observations in the earlier chapters of the book and in the last section. The most prominent among them is the existence of a smooth microwave background radiation with small but measurable temperature fluctuations. Some alternative models (like 'quasi-steady state cosmology') which are available in the literature approach the problem of structure formation from a very different angle. These models face serious difficulty in accounting for the cosmic microwave radiation. However, to be fair to such non-conventional models, it must be said that the possibilities within such models remain largely unexplored, since only a fraction of cosmologists work on them. In this book, we have followed the conventional thinking of the majority of cosmologists, and hence we will assume that the division between a smooth background and structures grown out of gravitational instability is a useful paradigm.

If we accept the above philosophy, then we could ask how well observations support either part of the paradigm. As regards the smooth background cosmology, the situation seems to be fairly secure. There are no concrete observational results which contradict any of the features of the background cosmology. It is, of course, possible to rule out a *subset* of models using these observations; for example, the age of the oldest stars in our galaxy puts constraints on the matter density and Hubble constant of the universe. Every once in a while, observations seem to indicate some serious problem (like the age of the universe coming out to be less than that of the oldest stars!) and the popular press has a field day. But invariably, such results either rule out only specific models or turn out to be false alarms. The observations could easily lead to a situation in the future in which only a very narrow range of parameters are viable for the background cosmology. If this happens, it still would not spell disaster for the big bang model but only to certain theoretical prejudices regarding the values of various cosmological parameters. For example, during the early and mid-1980s, several cosmologists strongly pushed for a universe with the matter density equal to the critical density. Such a value for the matter density was strongly favoured by certain theoretical speculations (like inflation). If the actual matter density observed in the universe turns out to be not equal to the critical density, it might rule out some of these theoretical speculations but it does not mean that the big bang model is in error.

On the positive side, the big bang model provides the most natural scenario for the thermal nature of the microwave background radiation and is consistent with the observed abundance of primordial helium and deuterium. It also provides a convenient framework to study several other observed high redshift phenomena, and has been reasonably successful in this task. For these reasons, most cosmologists would consider the big bang model as fairly well established, though they will differ drastically in their opinion about the values of the cosmological parameters like, say, the Hubble constant or matter density.

The situation is far less clear as regards the models for structure formation. To begin with, almost all cosmologists will agree that structure formation took place via gravitational instability. This is possibly the most natural mechanism to form structures within the context of big bang cosmology, and is an almost inevitable consequence of accepting the division between the background universe and structures. The anisotropies seen by COBE in the microwave background radiation provide a strong support for this scenario.

Given the gravitational instability picture, there are three ingredients which determine the details of structure formation. The first is the nature of the dark matter present in the universe. The second is the mechanism by which initial density fluctuations were generated. The third and last is the detailed hydrodynamical processes which operate in the non-linear phase of the gravitational instability. Unfortunately there exist serious uncertainties as regards all the three components mentioned above.

While most astrophysicists will agree that the dominant component of dark matter is made of weakly interacting massive particles, nobody has any real clue as to what these particles are. (For this reason there is less support for WIMPs among cosmologists than compared to, say, the big bang model; but it seems fair to say that the majority of them believes in WIMPs.) What is probably worse, it is not quite clear how one could go about checking the properties of a hypothetical particle like the WIMP. Particle physics models which predict these are not sufficiently well developed to give clear experimental signatures which are testable in the laboratory. Until a dark matter candidate particle is actually seen in the laboratories, the nature of the dark matter will plague cosmological model building.

The second aspect, namely, the origin of density fluctuations, is also fairly uncertain. The most successful model in this regard was the one based on inflation. In such a model inherent quantum fluctuations at microscopic scales give rise to density fluctuations in the universe at a later stage. Given a particular model for inflation, one can predict the kind of density fluctuation generated at the initial moment. In principle, this prediction can be tested by measuring the anisotropies of the microwave background radiation; in practice, however, this is not easy because of the observational uncertainties. This situation could improve in a decade or so if better quality observations become available. There is, however, another complication. Inflation is only one among several possible ways of generating the initial density fluctuations. It is possible to have other scenarios leading to the same predictions. From the point of view of structure formation, this is probably irrelevant because we only need to know the details of the initial density fluctuations and not the actual process which generated them. But, from a point of view of fundamental understanding, one would very much like to have the full picture.

Given the initial density fluctuations and the nature of dark matter, one can make fair amount of progress in models of structure formation as long as the density fluctuations are small. The next level of difficulty arises when gravitational instability enters the non-linear phase and hydrodynamical processes become relevant. Modelling complex hydrodynamical processes is a challenge to even the fastest computers available today, and hence progress has been somewhat limited. This situation, however, is likely to improve within the next decade or so, and hence it is only a question of time before more accurate predictions can be made, based on any structure formation models.

The key issue, of course, is whether *any* such model will eventually be able to account for *all* cosmological observations. In this regard the history of structure formation has been a history of near misses and glorious failures. The first model to be studied seriously was the one based on hot dark matter, which we now know is not viable. The second one was based on a cold dark matter scenario and held out hopes in the mid-1980s. This model, when it was originally proposed, had very few free parameters and possessed a tremendous amount of predictive power. Over the years, it was clear that the most natural version of such models cannot account for all observations, and people started introducing extra parameters into the theory. The COBE observations in 1992 completely ruled out the original cold dark matter scenario. In the post-COBE years, most cosmologists tried to create variations of the original model which had better chances of success. The models with mixed dark matter or with a cosmological constant – discussed in chapter 6 – belong to this class. Of these models, it is doubtful whether mixed dark matter models can reproduce an abundance of structures like damped Lyman-α clouds (see chapter 7, section 5). The model with the cosmological constant fares better but still lives rather dangerously. Strictly speaking, we can therefore say that there *does* exists at least one model (a variant of the cold dark matter model with a cosmological constant) which is not in contradiction with any observed phenomena.

The discussion in the above paragraph was purely pragmatic in order to bring the issues into sharp focus. Theoretical sciences, however, have benefited tremendously from aesthetic considerations, and one should not completely ignore aesthetic aspects of a model, however subjective such criteria can be. It has been generally accepted by scientists that models with fewer parameters and which make predictions are better models. From this criteria, both hot and cold dark matter models – in their original form – were very good models: they had very few free parameters which one could adjust and they made several predictions. Compared to this, models which have been in vogue after 1992 have been aesthetically quite unpleasing. In these models, one intro-duced parameters such that some observation is 'explained' post facto rather than for any fundamental reason. For example, there is no reason for the cosmological constant to have a non-zero value in the present universe, but the only viable model we have for structure formation today will work only if this parameter is postulated to have a very specific value! While most cosmologists agree that these models are not as 'nice' as the original models, they do not feel such aesthetic considerations are strong enough reason to stop working on such models. It is generally hoped that these issues are related to some aspects of physics which are as yet unknown and will get clarified at a later date.

While on the topic of aesthetics, one cannot ignore the issue that classical big bang cosmology predicts a singularity at some finite time in the past. This is probably the most unaesthetic feature any theory can have! Of course, we expect the theory to break down at very high energies (see chapter 5) and we hope that it will be replaced by a better model free of such singularities. Such a correct model might have some bearing on the issues of structure formation which we cannot even speculate upon at present.

The subject of structure formation has been one of the most active research areas in the last two decades or so. It had notable theoretical and observational triumphs and some disastrous failures. While scientists working in this area might differ on their perceptions on successes and failures, they all would agree that these have been very exciting years for the history of cosmology. One can be fairly certain that this excitement will hold sway in the years to come.

Glossary

21 cm line: In a neutral hydrogen atom, the spins of the electron and proton can be either aligned in the same direction or opposite to each other. These two states differ slightly in energy. When the atom makes a transition from one of these states to the other, it emits a photon with a wavelength of 21 cm. This radiation has been used extensively by radio astronomers to detect neutral hydrogen.

Å: see *Angstrom*.

Absorption (of radiation): When radiation passes through a material medium, its energy goes into exciting the atoms or molecules to higher energy states. This process is called absorption of radiation.

Accretion disc: A massive gravitating body (for example, a neutron star) attracts the surrounding gaseous material towards it. This material forms a disc-like structure around the central object, called an accretion disc. These discs play a crucial role in the modelling of active galactic nuclei and quasars.

Active galactic nuclei: Some galaxies emit significantly more radiation than the average ones. It is conjectured that this energy emission is due to the existence of some special processes near the centres of these galaxies. These objects are called active galactic nuclei.

Angstrom: One hundred-millionth of a centimeter (10^{-8} cm). This unit of length is denoted by the symbol Å.

Antiparticle: A particle with the same mass and spin as another particle but with equal and opposite electric charge. For example, the positron is the antiparticle of the electron; it has the same mass as the electron but is positively charged.

BL Lac objects: Immensely compact celestial objects with a vast but variable output of radiation. They belong to the set of *active galactic nuclei* and show similarities to quasars.

Beta decay: The transformation of a neutron into a proton, an electron and an anti-neutrino. This process is caused by the weak nuclear force and leads to a diminishing of the number of neutrons vis-à-vis protons in the early universe.

Bias: In the context of galaxy formation, this is a parameter which denotes how strongly clustered visible galaxies are compared to the underlying dark matter. A bias of factor 2 will mean that visible galaxies are clustered together twice as strongly as dark matter particles.

Binary (or binary star): A pair of stars in orbit around each other.

Binding energy: is the energy needed to separate a system into its constituents. For example, the hydrogen atom can be broken apart into a proton and an electron by supplying 13.6 eV of energy; so we say that the *atomic* binding energy of hydrogen is 13.6 eV.

Black body radiation (or black body spectrum): The characteristic radiation emitted by a body in thermal equilibrium at some given temperature; also called the Planck spectrum or thermal spectrum.

Blackhole: A region of space containing an immense concentration of material, leading to such a strong gravitational field that matter and energy cannot escape from that region.

Blueshift: A shift of the lines in a spectrum towards shorter wavelengths. Such a shift can occur if, for example, the radiating object is moving towards us.

Bosons: Sub-atomic particles like photons, gluons, W, Z, etc. Some of these particles play a key role in mediating interactions between other particles.

CDM: stands for Cold Dark Matter. This term refers to a particular class of elementary particles which could make up the bulk of the dark matter present in the universe.

CMBR: stands for Cosmic Microwave Background Radiation. Observations show that our universe behaves as though it is embedded in a box at a temperature of about 2.7 K. This radiation, believed to be a relic of an earlier hot phase of the universe, is called the cosmic microwave background radiation.

COBE: stands for Cosmic Background Explorer. This satellite was launched to study background radiation in the universe, and detected the temperature fluctuation in the CMBR in 1992.

Calorie: A unit of energy equal to 4.2×10^7 ergs. It takes about 80 calories to convert 1 gram of ice into water.

Chandrasekhar mass: is the limiting mass for a neutron star, and is about 1.4 times the mass of the Sun. Stars which are more massive can end up as black holes.

Correlation function (angular and spatial): is a mathematical function which characterizes the clustering of galaxies.

Cosmological constant: refers to a term in Einstein's equations describing gravity. If it is non-zero, it can help in solving some of the problems faced by structure formation models.

Critical density: The minimum energy density which is needed for the expansion of the universe to be halted eventually and be followed by a contraction phase.

Dark matter: Matter which is not visible in optical or other wavelengths but is detectable by its gravitational effects. There is evidence that our universe contains nearly ten times more dark matter than visible matter.

Decoupling epoch (or recombination epoch): A period nearly 400 000 years after the big bang when the temperature of the universe was low enough to allow the formation of neutral atoms. Radiation decoupled from matter around this epoch.

Degeneracy pressure: is the pressure exerted by particles (like electrons or neutrons) because of their quantum-mechanical nature, when they are compressed to high densities. This pressure plays a vital role in the late stages of stellar evolution.

Density contrast (or inhomogeneity): refers to the relative variation in the density of matter in the universe. Larger density contrast implies that the matter distribution is less smooth.

Density wave theory: is a model explaining the spiral structure of galaxies.

Deuterium: is an atom made of one proton and one neutron in its nucleus with an electron orbiting around it. This structure is similar to that of hydrogen except for an additional neutron in the nucleus. The amount of deuterium synthesized in the early universe puts constraints on the density of normal matter in the universe.

Doppler shift (or effect): refers to the shift in the lines of the spectrum towards the red or blue end caused by the motion of the source emitting the radiation. Redshift will occur if the source is moving away from us and blueshift if the source is moving towards us.

Electron: a fundamental, stable, atomic particle belonging to a class known as leptons. It is negatively charged and has a mass of about 0.5 MeV.

Electron volt (eV): is a unit of energy used in atomic physics. One electron volt is 1.6×10^{-12} ergs. One kilo electron volt (keV) = 1000 eV; 1 million electron volt (1 MeV) = 10^6 eV; 1 giga electron volt (1 GeV) = 10^9 eV.

Ellipticals: A major class of galaxies, ellipsoidal in shape.

Emission (of radiation): occurs when an atomic or molecular system makes a transition from a higher energy state to a lower energy state.

Energy levels: According to quantum theory, an electron bound in the hydrogen atom (say) cannot have an arbitrary value for its energy. The allowed values for energies for the electron are called the energy levels. In general, the allowed values of energy for any molecular or atomic system is denoted by this term.

Erg: The basic unit of energy. This is the kinetic energy of a particle of mass one gram moving with a speed of one centimeter per second.

Escape speed: The speed with which an object must be thrown into space in order to escape from the gravitational attraction of a celestial body. For an object thrown from the Earth, the escape speed is about 12 km s^{-1}. That is, if a body is thrown with a speed higher than 12 km s^{-1}, it will escape from the gravitational attraction of the Earth.

eV: see *Electron volt*.

Galaxy: A collection of (typically) 10^{11} stars which is gravitationally bound together. This is the basic 'unit' of matter in cosmology.

GeV: see *Electron volt*.

General relativity: refers to Einstein's theory of gravity. When the gravitational field is very strong, it is correctly described by Einstein's theory rather than Newton's law of gravitation.

Globular cluster: A closely packed cluster of stars which is spherical in shape. Such clusters contain anything from a few tens of thousands of stars to over one million. Globular clusters form a major component of the halos of galaxies.

Gluon: A particle which mediates the strong nuclear force and holds together quarks.

Gravitational lens: Massive bodies can make the path of light rays curve and hence can act in a manner similar to a lens. This effect can lead to the appearance of distorted or multiple images of distant celestial objects.

Graviton: A hypothetical particle which mediates gravitational interaction.

HDM: stands for Hot Dark Matter. This term refers to a particular class of elementary particles which could make up the bulk of the dark matter present in the universe.

H–R diagram: Stands for Hertzsprung–Russell diagram. A graph in which the luminosity of a star is plotted against its temperature in some suitable manner.

Hadron: A class of sub-atomic particle which experiences the strong nuclear force; made of quarks.

Helium: is made of two protons and two neutrons in the nucleus and two electrons orbiting around it. This is the second most abundant element in the universe after hydrogen.

Hertz: is a unit of frequency corresponding to one oscillation per second.

Hubble constant: Refers to the rate at which the universe is expanding. Observationally, it has a value between 50 km s^{-1} Mpc^{-1} and 100 km s^{-1} Mpc^{-1}.

Hydrogen: The lightest and most abundant chemical element; made of a proton and an electron.

Inflationary model (of the universe): refers to a model for the universe in which a very rapid expansion took place fairly early in the evolution of the universe.

Intergalactic medium: refers to the material that exists between galaxies. There is evidence to show that the intergalactic medium is ionized.

Interstellar medium: refers to the material that exists between stars in a given galaxy.

Inverse square law (of force): Gravitational and electromagnetic forces become weaker by a factor of 4 if the distance is increased by a factor of 2, and by a factor of 9 if the distance is increased by a factor of 3, etc. Such a force is said to obey the 'inverse-square' law.

Ionization: is the process by which electrons are removed from atoms.

Isotopes: are nuclei with the same number of protons but a different number of neutrons. For example, hydrogen (which has one proton in the nucleus and zero neutrons) and deuterium (which has one proton and one neutron in the nucleus) are isotopes.

keV: see *Electron volt*.

kpc: see *Parsec*.

Kelvin: is a unit of temperature such that zero degrees kelvin refers to about -273 degrees centigrade. To obtain the temperature in kelvin, one should add 273 to the temperature expressed in centigrade; for example, 27 C corresponds to 300 K.

Lepton: A class of sub-atomic particles which are not affected by strong nuclear forces. There are six known leptons, viz. electron, muon, tau lepton and the three types of neutrino.

Light year: Distance travelled by light in one year; about 9.5×10^{12} km.

Lyman alpha line: is the radiation emitted when the electron jumps from the first excited state to the ground state in the hydrogen atom.

MBR: see *CMBR*.

Magnitude (of stars, galaxies): refers to a way of characterizing how bright a celestial body is. The stars of first magnitude are about 2.5 times brighter than those of second magnitude, etc.

Main sequence (stars): are those which shine essentially by nuclear fusion of hydrogen into helium. In the H–R diagram these stars form a band from top left to bottom right.

MeV: see *Electron volt*.

Mpc: see *Parsec*.

Muon: is a sub-atomic particle belonging to the family of leptons.

Neutrino: is a sub-atomic particle belonging to the class of leptons. There are three kinds of neutrinos, called the electron-neutrino, muon–neutrino and tau-neutrino. Their masses are uncertain; if massive, they can be viable candidates for HDM.

Neutron: sub-atomic particle with a mass of about 939.6 MeV. It carries no electric charge and is a constituent of the atomic nucleus.

Nucleosynthesis: refers to the processes by which elements heavier than hydrogen are formed by nuclear reactions. Very early in the evolution of the universe, such nucleosynthesis occurred, leading to the formation of (mainly) helium and deuterium. All other heavier elements are formed inside the stars.

pc: see *Parsec*.

Parsec (pc): A unit of distance approximately equal to 3×10^{18} cm; 1 kiloparsec (kpc) $= 10^3$ pc; 1 megaparsec (Mpc) $= 10^6$ pc.

Photon: is the quantum of electromagnetic radiation which mediates the electromagnetic forces between charged particles.

Planck spectrum: see *Black body radiation*.

Plasma: An ionized gas made of electrons and ions moving freely and randomly.

Population I and II (stars): Population I stars are young, later generation stars found in the spiral arms of our galaxy. Population II stars are older, earlier generation stars with a smaller fraction of heavy elements.

Proton: is a sub-atomic particle with a mass of about 938.3 MeV. It is made of three quarks and carries a positive electric charge. It is a constituent of all atomic nuclei.

Quark: is a fundamental particle which is the constituent of all hadrons. There are six kinds of quarks, called 'up', 'down', 'top', 'bottom', 'charmed' and 'strange' quark. The neutron, for example, is made of one up and two down quarks.

Quasar: is an acronym for Quasi-Stellar Radio source. They have high redshifts, star-like appearance and emit almost as much radiation as a galaxy.

Redshift: refers to the shift in the spectral lines towards longer wavelengths. When this occurs due to the motion of the source, it is called the Doppler shift, and when it arises due to the gravitational field of the emitting object, it is called gravitational redshift.

Seyfert: A spiral or barred galaxy with an intensely bright central region. They show similarities to quasars, and are considered to be related to active galactic nuclei.

Singularity: In the context of cosmology, this refers to the moment at which the universe had zero size and infinite density. This is a mathematical artifact indicating the breakdown of the theoretical model.

Solar mass ($M_\odot$): is a convenient unit for measuring the masses of celestial bodies; roughly equal to 2×10^{33} gm.

Spiral: A galaxy that has a central bulge and spiral arms in its equatorial plane.

Supernova: Some stars undergo an explosion ejecting most of their material into space during the late stages of their evolution. Such an exploding star is called a supernova.

Tau: A sub-atomic particle belonging to the class of leptons.

Thermal radiation: see *Black body radiation*.

Uncertainty principle: states that it is not possible to measure the position and speed of a quantum mechanical particle with arbitrary accuracy at the same time. This principle forms the cornerstone of quantum theory.

WIMP: is an acronym for Weakly Interacting Massive Particle. These are hypothetical particles which might make up a large fraction of the dark matter in the universe.

Further reading

The following books are almost at the same level as the current one and discuss issues related to structure formation with different levels of emphasis:

1. S. Weinberg, *The First Three Minutes*, (New York: Bantam), 1977.

   I consider this book a model for popular science writing. Definitely worth reading, even after nearly 20 years since its initial appearance. A classic by any criterion.

2. Joseph Silk, *A Short History of the Universe*, (Freeman and Co.), 1994.

   Describes much of the same ground as covered in the present book but with a lot more pictures and colour photographs.

3. J. V. Narlikar, *The Structure of the Universe*, (Oxford University Press), 1980.

   A delightful book written in a very lucid, non-mathematical style.

4. J. Audouze and G. Israel (Editors), *The Cambridge Atlas of Astronomy*, (Cambridge University Press), 1994.

   Provides a colourful and comprehensive coverage of every aspect of astronomy. Strongly recommended.

5. G. Wynn-Williams, *The Fullness of Space*, (Cambridge University Press), 1992.

   This is a more focused book, dealing with the interstellar and intergalactic medium and, in that sense, somewhat limited in scope compared to the current book. But it provides a very good grounding in physics for the non-mathematical reader and contains valuable insights into astrophysics.

6. J. Gribbin and M. J. Rees, *Cosmic Coincidences: Dark matter, Mankind and Anthropic Principles*, (New York: Bantam), 1989.

   Somewhat outdated as regards current developments. However, as suggested by the title, covers a much broader range of topics.

7. P. Davies, *God and the New Physics*, (New York: Simon and Schuster), 1985.

   If you are worried about the wider implications of physics and cosmology, this is probably a good no-nonsense book to start with. Far superior to many other 'philosophical' books on the market.

The following books are intended for a mathematically sophisticated reader:

1. F. Shu, *The Physical Universe*, (University Science Books), 1982.

   A delightful introduction to the physical processes in the universe at all scales from planetary systems to cosmology. Requires very little mathematical sophistication, but more than is required for the current book.

2. J. V. Narlikar, *Introduction to Cosmology*, (second edition), (Cambridge University Press), 1993.

   Good introductory textbook in cosmology. Requires a fair amount of mathematical sophistication on the part of the reader.

3. T. Padmanabhan, *Structure Formation in the Universe*, (Cambridge University Press), 1993.

   Being more honest than modest, I will term this book as 'excellent'. Contains an up-to-date mathematical introduction to this subject.

Index